Beekeeping

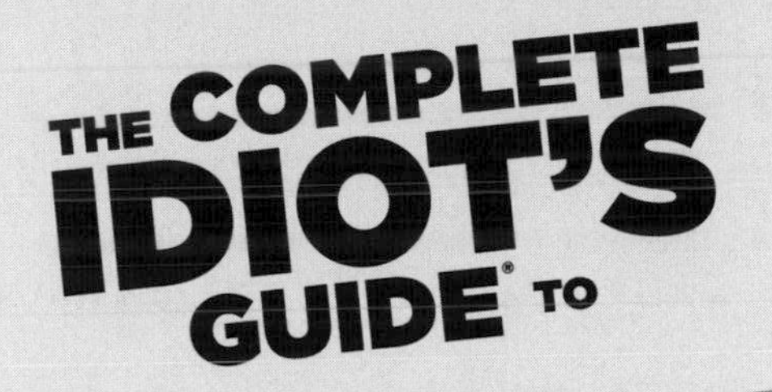

Beekeeping

by Dean Stiglitz and Laurie Herboldsheimer

A member of Penguin Group (USA) Inc.

ALPHA BOOKS

Published by the Penguin Group

Penguin Group (USA) Inc., 375 Hudson Street, New York, New York 10014, USA

Penguin Group (Canada), 90 Eglinton Avenue East, Suite 700, Toronto, Ontario M4P 2Y3, Canada (a division of Pearson Penguin Canada Inc.)

Penguin Books Ltd., 80 Strand, London WC2R 0RL, England

Penguin Ireland, 25 St. Stephen's Green, Dublin 2, Ireland (a division of Penguin Books Ltd.)

Penguin Group (Australia), 250 Camberwell Road, Camberwell, Victoria 3124, Australia (a division of Pearson Australia Group Pty. Ltd.)

Penguin Books India Pvt. Ltd., 11 Community Centre, Panchsheel Park, New Delhi—110 017, India

Penguin Group (NZ), 67 Apollo Drive, Rosedale, North Shore, Auckland 1311, New Zealand (a division of Pearson New Zealand Ltd.)

Penguin Books (South Africa) (Pty.) Ltd., 24 Sturdee Avenue, Rosebank, Johannesburg 2196, South Africa

Penguin Books Ltd., Registered Offices: 80 Strand, London WC2R 0RL, England

International Standard Book Number: 978-1-61564-011-9
Library of Congress Catalog Card Number: 2009938588

13 12 11 8 7 6 5

Interpretation of the printing code: The rightmost number of the first series of numbers is the year of the book's printing; the rightmost number of the second series of numbers is the number of the book's printing. For example, a printing code of 10-1 shows that the first printing occurred in 2010.

Printed in the United States of America

Publisher: *Marie Butler-Knight*
Editorial Director: *Mike Sanders*
Senior Managing Editor: *Billy Fields*
Acquisitions Editor: *Karyn Gerhard*
Development Editor: *Jennifer Moore*
Senior Production Editor: *Megan Douglass*
Copy Editor: *Emily Garner*
Cover Designer: *Kurt Owens*
Book Designer: *Trina Wurst*
Indexer: *Tonya Heard*
Layout: *Brian Massey*
Proofreader: *John Etchison*

Like everyone, we have parents, heroes, and mentors who have helped craft who we are and influence everything we do.

This particular book, however, could only be dedicated to Dee Lusby. The contributions that she and her late husband Ed have made and shared freely with many thousands of beekeepers all over the world are the driving force behind the treatment-free movement, and their value is impossible to overestimate.

Contents at a Glance

Contents

Introduction

One can imagine the progression of humans' relationship with the honeybee. Curiosity led to stinging, which drove us away. More curiosity led to the discovery of honey and, suddenly, the fear of being stung was no longer an effective deterrent. Humanity had never experienced such sweetness before, and the taste of liquid sunlight changed the world. It's quite literally a version of the oldest story in the book (think: Adam, Eve, and a sweet, forbidden apple).

Honeybees are the most studied creatures on the planet, second only to humans. Someone was the first to discover that smoke would drive bees out of their cavity and minimize alarm, what time of year there was likely to be the most honey, and that breathing on bees tends to rile them up. Such knowledge is, of course, the result of observation and study. But the bees don't make our research easy.

Gregor Mendel, whose discoveries described the genetic crossings of virtually every life form on the planet, was baffled by the honeybee. Even today, we are just beginning to understand certain aspects of the hive, such as the importance of the microbial components.

No matter the accumulated knowledge, no matter the sophistication of the tools, every answer we get yields countless questions. The closer we look, the deeper we go, and it never seems to end. This is the nature of the honeybee.

Working with them, studying them, trying to unlock their secrets, is both satisfying and engaging. This is the big secret: bees are more interesting than honey.

Don't get us wrong, honey is great stuff, and there is nothing like sticking your finger into warm honeycomb and having a taste while bees fly around you. Not even a cherry tomato warmed by the sun and right off the vine compares. However, we love the bees most of all, and "bee fever" is a constantly recurring theme in the history of humankind. We know we are not alone. Welcome.

Unfortunately, beekeeping has suffered—along with all of agriculture—from the introduction of modern high-yield techniques and industrial-scale approaches. Think about monocrop farming, feedlot cattle, chickens with their beaks removed, and excessive and improper application of pesticides. Similarly, many of our attempts to manipulate the natural processes of the bees have been careless and misguided, and we have treated Mother Nature's generosity with greed. We should know better.

Modern beekeeping practices have delivered significant challenges, and this book attempts to address them. Beekeeping by recipe doesn't really work unless you employ artificial controls such as frequent feeding, chemical applications, and constant requeening. Even with these interventions, it's not unheard of for 90 percent of a beekeeper's hives to die over the winter. Such solutions do not solve anything.

This is likely the first time you've ever heard that most beekeepers actually put chemicals in their beehives, or that beekeepers (yes, even the sweet old man at the end of the dirt road who sells honey out of his garage) feed their bees sugar or high-fructose corn syrup. It's shocking that beekeepers don't question these practices, and that some treatments are so common that beekeepers don't even consider them treatments anymore—they've become baseline beekeeping.

In addition to serving as an introduction to beekeeping, this book is intended for beekeepers of all levels of experience who want to keep bees on a system that is different than how most people are taught today. What we offer here is a treatment-free approach. We say "treatment-free" because even "organic" and "natural" beekeeping (as they have come to be defined) allow for chemical treatments and management practices that are detrimental to the long-term health and vitality of the honeybee.

We don't pretend to present a recipe. What we hope to convey is an understanding of how bees live, some overall management techniques and goals, and an overview of how bees function in the natural world so you can develop a mutually beneficial relationship with them.

A beekeeping recipe is the fish that one gives a man instead of teaching him to catch his own. We endeavor to impart to you the knowledge you need to create your own management approaches based on what the bees need at any given time.

How This Book Is Organized

To help you assimilate such a complex assembly of information, ideas, and practices, this book is divided into four parts.

Part 1, "What's All the Buzz About?" will introduce you to the bees, their fascinating lives both as individuals and members of the dynamic colony, and the microbes they can't live without. You'll learn a lot of new words, some bee biology, tour a hive, and figure out what equipment and supplies you'll need to get started with your own bees.

In **Part 2, "A Bee of One's Own,"** you'll discover where to find bees, how to get them, and what to do with them when they arrive. You'll learn how to select locations, make sure the bees get the food they need, and what to do as the colony begins to expand.

Part 3, "Beekeeping the Old-Fashioned Way," explores a treatment-free management approach. You'll learn about the importance of cell size and unlimited broodnest and how microbes impact the hive both in sickness and in health. You'll get a crash course in honeybee sex, understand how breeding can make or break honeybee populations, and learn how to replace a queen when circumstance demands it.

In **Part 4, "Harvesting and Beyond,"** you'll learn when and how to harvest honey, get some ideas for satisfying your bee cravings during the offseason, and start to make plans for growing your beekeeping operation. Opportunities for expanding your bee knowledge abound. Before you know it, it will be spring again—time for you to share what you've learned, and your bees, with others.

Extras

Throughout the chapters you'll see four different types of sidebars.

def•i•ni•tion

These sidebars introduce you to new vocabulary words. All of these terms, plus many others that you'll find in the text, can also be found in the glossary in Appendix A.

Bee Aware

Turn to these sidebars for items of interest, including historical, unusual, and almost unbelievable bee and beekeeping facts.

Bee Bonus

Here you'll find cautions and things to look out for.

Bee Smart

Reading these tips and pointers will help you become a better beekeeper.

Last but not least, we have created a website to support our readers: www.TheCompleteIdiotsGuideToBeekeeping.com contains additional information, updates, clarifications, educational videos, commentary from other beekeepers, and an interactive forum so that readers can communicate with one another—and with us. We hope to see you there!

Acknowledgments

There are of course, countless people to thank, and little room to do so. We will limit ourselves here to thanking those whose influence had an immediate and direct bearing on this book.

To Kirk Webster, Michael Palmer, Michael Bush, Erik Osterlund, Markus Barmann, Sam Comfort, Randy Quinn, Kerstin Ebbersten, and Dee Lusby for providing us with countless hours of education, obsessive honeybee conversation, speculation, innovation, and clarification.

Maryann Frazier, Jerry Hayes, Martha Gilliam, and Tom Gammell were key in leading us to, and helping us perform, the experiment we write about in "No Bee Is an Island," (BeeUntoOthers.com/NoBeeIsAnIsland.pdf) which has been pivotal in our understanding and appreciation of the microbial ecosystem within the hive.

We would also like to thank our parents, Glenn Stillman (along with his entire family and crew), Bruce Larson, Adam Stark, Debra Stark (and her crew), Jake Heinemann (and the rest of the Maxant crew), Christy Hemenway, Jim Phelan, Connie Richardson, Jimmy Xarras (and his crew), Matt Diprizio, Worcester County Beekeepers Association members and board, the city of Leominster, Dan Conlon and Mass Bee, Sovereign Bank of Leominster, and our friends at The Trustees of Reservations and the Federation of Massachusetts Farmer's Markets. Without their help, the bees could not support us.

We appreciate all who host our bees, and their neighbors, for providing an environment in which our bees can thrive.

We must also thank our retail stores, honey customers, conference and bee club meeting attendees and staff, friends, family, Internet buddies, and researchers (past, present, and future) who have been invaluable in supporting us and stimulating us to refine our ideas.

Somehow, a pipe dream of writing a book nearly effortlessly transformed into opportunity and then into reality. Our agent, Marilyn Allen of the Allen O'Shea Literary Agency, and our editors, Karyn Gerhard and Jennifer Moore of Penguin and Alpha Books, are responsible for finding us and initiating us into the world of publishing. We are indebted to them.

We hope the rest of you will forgive us for not naming you specifically. Know that we appreciate everything you have done for us. Your contributions have not gone unnoticed.

Trademarks

All terms mentioned in this book that are known to be or are suspected of being trademarks or service marks have been appropriately capitalized. Alpha Books and Penguin Group (USA) Inc. cannot attest to the accuracy of this information. Use of a term in this book should not be regarded as affecting the validity of any trademark or service mark.

Part 1

What's All the Buzz About?

This part is all about preparation. Yes, we know, prep work is usually boring. Trust us, when it comes to bees, everything becomes magically interesting. You're going to have almost as much fun learning about your future bees as you'll have when they finally arrive.

We begin with a crash course in vocabulary and virtual tours of the combs and the hive. You'll learn how the different members of the colony function, both as individuals and as a group, and who's really in charge. (Hint: it may not be who you think.) You'll also learn about the fascinating but underappreciated role that microbes play in the honeybee colony.

Preparation means thinking ahead, so you'll find out about where and how to keep bees and potential problems you may encounter. You'll have decisions to make about what equipment to purchase, tools you'll need, and the ever-burning question of what to wear!

We know you can't wait to get started, but believe us, when that box of bees shows up you'll be really glad that you took the time to get ready!

Chapter 1

The World of Bees

In This Chapter

- The benefits of bees
- Learning beekeeping lingo
- Taking a closer look at a brood frame
- Touring a virtual hive
- The many kinds of beekeepers

Congratulations! You are about to embark on a life-changing course. We know this may sound dramatic, but deciding to become a beekeeper truly is a pivotal event for all of us who have bettered our lives with bees. This is because bees are, simply put, completely amazing. By getting up close and personal with them on a regular basis, you'll come to know on ever-deepening levels all that the bees have to offer. We promise you won't regret it.

The laws of nature are reflected in the culture of the honeybee perhaps more clearly than they are in human culture. Traditionally, some obvious anthropomorphic qualities have been ascribed to honeybees: monarchy (in the mistaken belief that the colony is "led" by the queen), and socialism

> **Bee Bonus**
>
> Wild honeybees and managed honeybees are both known by the same elegant Latin name, *Apis mellifera*. There is virtually no difference between the two.

(the individual worker serves the whole of the colony selflessly). To the astute observer and the curious mind, the behavior of the honeybee illustrates fundamental principles of economics, engineering, population dynamics, manufacturing, and thermodynamics! The honeybee offers an entire curriculum of study observable inside a wooden box.

The Needs of the Bees Are Simple

Keeping bees requires little more than providing them with a simple shelter in which to create their comb, store their honey, and raise their young. For a few hundred dollars, anyone can set up a hive or two and get started. Bees on a natural system require very little from you. The bees manage their own population and gather and produce their own food without any help.

The bees manage their own reproduction on both the individual level—producing more bees—and on the colony level—swarming (which is not a stinging attack, but a completely nonaggressive act of reproduction). They have the resources to severely impact predators by stinging, yet allow beekeepers regular access to their hive's inner sanctum. They produce enough food to both survive the winter and continue as a colony. And most years, they can produce enough extra honey to share with the beekeeper.

> **Bee Bonus**
>
> Bees tell each other about the direction, distance, and quality of available forage through a dance language that is performed on the wax comb inside the hive. Because these dances are executed in the dark, the bees feel the vibrating comb through their six feet and translate the information into a flight plan, using the angle of the sun to orient.

If you have a few square feet of outdoor space to devote to a hive, you can become a beekeeper. A patio, a city rooftop or balcony, small backyard, big backyard, rural field, farm, or orchard all are fine places to keep bees. Because bees fly to forage for their food, you do not have to provide the flowers for the pollen and nectar they collect. In fact, if you did, you would have to own a few square miles of land, as that's how far bees range in search of forage!

Once you provide housing for the bees, they will come and go as they please. Bees have an amazing sense of direction, and they almost always find their way home. Unlike most other livestock or pets, no fencing, enclosures, or leashes are required. Bees need no daily feedings or waste cleanup, no milking, no grooming, and no visits to the vet. In fact, short of a few times per year when bees may benefit from your attention, the more you let them do their thing uninterrupted, the happier and more productive they will be.

Bee Nature and History

Honeybees provide an almost unsurpassed opportunity to connect with nature. You've probably seen managed beehives—typically, stacks of wood boxes painted white—from afar, maybe on the edge of a field or a rooftop. If you got within a few feet of the hives during the day anytime between spring and late fall, you probably saw bees flying near the hive.

If you were brave enough to get close enough to touch the hive, you saw an abundance of activity: bees coming and going from the entrance with colorful pollen packed on their back legs, guard bees protecting the entrance from intruders, and maybe a more rare event, such as a virgin queen emerging from the hive to conduct her mating flight. If you pay attention, bees provide more drama than a TV miniseries—all for the price of a chair by the hive and as much time as you're willing to spend taking it all in.

Bee Aware

One of the first questions people ask about keeping bees is whether they will get stung. The short answer to this question is *yes*. You probably will get stung if you spend much time with bees. But unless you are allergic to bee stings, a sting now and then isn't really a big deal.

Beekeeping also connects you to very ancient times. Honeybees have been found preserved in amber from 100 million years ago, indicating that they survived the asteroid impact that likely wiped out the dinosaurs. There is evidence that our Paleolithic and Neolithic ancestors hunted for honey. Mesolithic paintings (perhaps from 11,000 years ago) in Spain depict humans climbing a vertical rock face and an unstable ladder in search of honey high up in the rocks. The paintings even show someone falling off the ladder, suggesting that humans haven't changed much in their willingness to take great risks for the right prize (and the unstable ladder remains part of the beekeeper's standard equipment)!

The intricate patterns revealed by the bee's construction and behavior are as attractive to the intellectual cravings of humanity as the honey is to our craving for sweets. Over thousands of years the arts of honey hunting and beekeeping advanced along with our understanding of the bees themselves. Scientists, philosophers, enthusiastic amateurs, and professional beekeepers have all contributed to the body of knowledge that is the foundation of what we know about bees today.

Modern beekeeping began in the mid-1800s with the introduction of the moveable frame hive. Little has changed since then, as the techniques and equipment developed at that time are so well suited to the keeping of bees.

What has changed in recent years is the shift from honey production to pollination as a main income source, and the use of treatments inside the hive to combat threats new and old. In essence, it's the same set of mistakes made in all of agriculture: increasing production is demanded from the same set of resources year after year. This approach will always break down eventually, and it's precisely why we offer an alternative.

An Introduction to Bee-Speak

In order to begin any sort of meaningful discussion about bees and beekeeping, you need to learn some beekeeping terminology. Don't worry, most of these terms are fairly straightforward, and if you forget the meaning of a term, all you need to do is turn to the glossary in Appendix A for a quick refresher:

- **Apiary** A yard with beehives on it; a bee farm.
- **Stand** A support that one or more hives sits on.
- **Colony or hive** A self-contained group of bees that live and work together. *Hive* can refer to the bees and the structure, or just the structure used to house the bees. A healthy colony of bees can contain between 15,000 and 100,000 individuals.
- **Hive body** Any one of the single boxes full of frames in a modular stack of boxes that makes up the hive and houses a single colony.
- **Super** A box that holds frames for honey production by the bees, also known as honey super. It is placed above the broodnest. Note that *hive body* and *super* are sometimes used interchangeably.
- **Brood** Young bees in every stage of development from the time the eggs are hatched to when the adult bee emerges.
- **Broodnest** The area in the combs where eggs are laid and brood is raised. The broodnest generally occupies the center of the hive over multiple combs.

- **Swarm** When some (not all) of the bees, including a queen and workers, leave the original colony to establish a new colony. Swarming is the reproduction of a colony into two colonies. Swarms have no resources to protect (comb, food stores, brood), and tend to be extremely passive and nonaggressive.
- **Comb** A sheet of horizontally oriented beeswax with hexagonal tubes that bees use for producing and storing honey, storing and fermenting pollen, rearing their brood, and living. Comb has two similar sides, and the sides are offset from one another for maximum strength.
- **Cell** One of the many individual hexagonal wax tubes that make up the comb.
- **Drone** The male honeybee. The drone is large, has no stinger, and is produced from an unfertilized egg (and therefore has no father). Drones have the longest gestation period (24 days) of all the castes, or types, of bees.
- **Queen** The reproductive female of the hive. She lays all the eggs, and is mother to all the bees born in the colony. Queens develop very quickly (probably due to the incredible amount of protein in their diet), and emerge 16 days after the egg is first laid. A newly emerged queen must first take orientation flights of increasing lengths so that she can learn how to get back to her colony. Eventually (a week or two after emerging) she will go on one or more mating flights to transform from a virgin (unmated) to a laying (mated) queen. During her mating flights she will mate with up to 30 drones. There is generally only one laying queen in the hive (although it isn't uncommon to find a mother and daughter both laying), and virgins will fight to the death with one another, and even sting not-yet emerged queens in their cells.
- **Worker** The vast majority of the bees in the hive. Workers are nonreproductive females who do all the work in the hive besides laying eggs. Workers (and queens) are both raised from fertilized eggs; the difference in development has to do with the diet each is fed. Workers develop from newly laid eggs into adults in 21 days.
- **Beebread** Fermenting pollen that is packed into cells. Beebread is the protein source for the entire colony, and the building material for raising new bees.
- **Nectar** The sugary substance produced by flowers in order to attract pollinators. Bees collect nectar and transform it into honey.
- **Honey** Produced from nectar by both evaporating out most of the water, and via enzymatic transformation by microbes living in the bees themselves. Honey is food/energy storage for the honeybee colony for both the long and short term.

- **Honey Stomach** (or "crop") is a sac separate from the bees' digestive tract where bees store nectar and where some of the transformation into honey takes place.

Up Close in the Broodnest

Now that you know some of the basic terminology, let's take a close look at what actually goes on in a beehive. The core of activity centers around the broodnest.

Bee Bonus

An observation hive (where you can see the bees working on comb behind glass) allows you to see the queen laying her eggs, witness the bees feeding the larvae, feeding and grooming the queen, building cells, evaporating nectar, heating the brood, cleaning the cells, and much more. The observation hive is an invaluable tool in understanding the honeybee.

A Spiral Womb

The queen typically lays eggs in a spiral pattern in the broodnest, starting from the center and working her way out. As she moves around the comb, the queen pokes her head into cells for inspection. If a cell seems promising she'll go into the cell up to her shoulders, turning around and inserting her abdomen to lay an egg against the back of the cell if it's just right.

The central worker brood is covered with tan-colored wax cappings (capped by adult workers nine days into development). The cappings are slightly raised and roughly textured.

In the middle of the brood, where the queen first began laying eggs, some cappings are cracking open and wiggling antennas are poking through the wax; worker bees are emerging. Over the next half hour or so, the bees will chew their way through the caps and begin to poke their heads through. With a final push, a whole bee will pop out ready to begin house duties.

Circling around the periphery of the capped brood are larvae in various stages of development, the largest of which are adjacent to the capped brood in the center. The smallest (youngest) of the larvae are barely visible; the biggest fill the cell from side to side. The larvae resemble glistening white shrimp curled in shimmery pools of brood food.

Capped worker brood.

At the outermost reaches of the broodnest are the eggs. Nearly invisible, they resemble tiny, translucent grains of rice.

Eggs attached to the back of the cell walls and larvae floating in brood food.

If you were to come back in a few days, you'd find the cells in the center where bees had emerged filled with eggs, and the entire process starting over, with the size of the active broodnest expanding and contracting with the season and the needs of the colony.

If there are drone cells, they will tend to be at the outside edges of the frame. Drone cells are bigger then worker cells and capped drone brood is bullet shaped, protruding out from the comb conspicuously.

Food Near the Brood

In cells near the brood, the bees store fermenting pollen (beebread) and honey. The pollen can be any number of colors and is slightly recessed in the cells. Uncapped honey may vary from light to dark and shines in the cells. When the bees have removed enough moisture from the honey, they build wax cappings over the top to seal it. Capped honey is covered with a thin, white layer of wax that is flatter and smoother than brood cappings.

Bee Bonus

Try as they might, humans have not been able to produce honey without bees. The latest research by Dr. Alejandra Vasquez and Dr. Tobias Olofsson in Sweden indicates that enzymes from unique bacteria in the bees are essential for the conversion of nectar to honey, which would classify honey as a fermented food.

The Staff

Because the brood comb has larvae in it, worker bees —primarily house bees and nurse bees—cover the brood to keep it warm and to feed the larvae. Worker bees do many tasks in and around the brood, some of which are difficult to decipher by merely watching. Much of their energy goes to processing nectar and pollen, caring for the brood, and preparing cells for the queen to lay eggs.

Her Royal Highness

Somewhere in the undulating mass of bees is the queen. Several physical attributes set the queen apart from the workers. Her *thorax* is larger than the workers', and the black spot on the back of her thorax is bigger and shinier. Her abdomen is much longer than the workers', making her wings appear much shorter in proportion to the rest of her body. As the queen only uses her wings for her orientation and mating flights or swarming, she keeps them tucked behind her back while she's in the hive.

def•i•ni•tion

The **thorax** is the middle section of the bee from which the wings and the six legs are attached.

The queen can often be seen resting on the comb, encircled by a group of workers who are feeding and grooming her. Otherwise, she marches deliberately around on the comb, poking her head into cells, looking for suitable places to lay eggs.

A queen surrounded by her worker attendants.

Protruding from the comb, hanging vertically with the openings facing down, may be queen cups. They resemble small acorn tops and are kept by the bees in case a new queen needs to be raised. If an egg or young larvae is put into the queen cup, and the bees decide to raise it, they will build a queen cell down from the cup. The cell resembles a peanut attached to the comb. The exterior texture of the queen cell is very interesting in that it has a honeycomb-ish pattern in miniature embossed in the wax.

The Drones

Drones are raised similar to workers, except they require more resources. They are larger than workers and are raised in larger cells. The 24-day gestation period for drones is longer than for workers or queens, and they consume the most food in development. Drones can't even chew through their own cappings to emerge and must instead rely on workers to release them. For the first three days of their lives, drones can't even feed themselves.

Bee Smart

New beekeepers often mistake the drone for the queen. Look at the eyes. If they meet at the top of the head, it is a drone, not a queen.

Besides its size (large) and shape (like a brick), the other noticeable feature of the drone is its eyes—they are huge and meet at the top of the head. The only real active job for the drone is to catch a queen in flight to mate, which is a relatively rare occurrence. Spotting the queen in flight requires such considerable optical equipment.

A drone and a number of workers. Note the size of the drone and his huge eyes.

A Virtual Tour of the Hive

Now that you have a better idea of what is going on in the broodnest, let's take a virtual tour of the hive. Since hives vary significantly throughout the year, the tour will consider the hive during each of the four seasons.

The Winter Cluster

If you live in northern climates, you may have to wade through some snow to get to the hive in the winter. Once you're there, you'll probably see some signs of heat in the form of melted snow around the hive. That's good. It means that the bees are alive and well.

Inside you'll see combs of wax cells. The bees are clustered together on the combs. The colder it is, the tighter the cluster. The bees may actually be situated head first inside the cells, one bee per cell, on opposite sides of the comb, sharing heat through the wax walls.

Cells are partially emptied of stored honey, the bees having already eaten part of their stores. The honey is generally stored across the upper corners of the brood frames or in full frames next to and above the broodnest.

In the center of the broodnest, there may already be a patch of brood. The bees have to prepare well in advance of spring so they have populations ready to bring in food when the first pollen is on the trees and early plants. If they were able to store a lot of pollen and honey before winter, they'll get a head start on spring. If it's a warmer winter day, bees may be moving around. They might even take a cleansing flight—a trip to the little bee's room—leaving yellow dots on the snow. The bottom board of the hive probably has some accumulated dead bees, as not all of them make it through the winter.

Spring Awakening

As spring unfolds and the first plants start producing pollen, the bees' seasonal work begins. In many ways, the coming of spring really means planning for next winter, as ultimately everything the bees do is to perpetuate the long-term survival of the queen's genetic line.

The bees begin bringing pollen into the hive and ramping up their brood rearing. The queen increases her egg laying, moving in spirals from one comb to the next, and the broodnest expands. Initially, the brood consists of all workers, but as the season progresses, the bees raise drone brood. Colonies can't reproduce or successfully replace their queen until drones are available to fertilize the virgin queens. Signs of drones mean that the reproductive season is underway.

Bee Bonus

Every geographical area has its unique climate, weather patterns, bloom patterns, available forage, and other variables. These inputs vary year to year because of both natural causes and human intervention. Bees have to be able to adapt quickly to these variables to ensure their survival.

Summer

As spring moves into summer, the hive population increases to take advantage of forage. The bees bring in pollen and nectar as they become available from plants and trees, and store and process them in the cells. The goal is to generate an overall surplus so the hive can make it through another winter and reproduce (swarm).

If there is a large surplus and they run out of space, the hive will build queen cells in preparation for swarming. This impulse can be redirected by the beekeeper into honey and/or bee production, and will be discussed throughout this book.

Fall

In some parts of the country, late summer is a time of increased bee activity. The last of the blooms are on plants and trees, and the bees prepare for winter by bringing in as many stores of pollen and nectar as possible.

As summer moves to fall, the workers kick most of the drones out of the hive. The drones' mating services will not be needed until spring, so they are generally not allowed to overwinter and use the resources of the hive. Struggling drones are seen trying to re-enter the hive, where they are promptly dragged out again by the workers and dropped on the ground to starve. The bees keep a few drones through the winter just in case their services are needed for early emerging queens.

In hives that do see winter, the amount of brood shrinks until all the brood is hatched. These last bees to emerge for the season don't use their resources feeding brood, and therefore have fatter bodies for better overwintering. If the colony has been able to bring in enough stores, full frames of honey flank the broodnest area. (By this time any extra honey supers that the beekeeper may have added to collect surplus honey have been removed.) The cells in the center of the broodnest should be open so that the bees have space to cluster and the queen has room to lay when the time comes. Bees vary in the size of clusters they need to overwinter. Some breeds of bees overwinter with very small clusters and require fewer stores to make it to spring.

At some point in the winter, the bees anticipate the early pollen and nectar flow and the cycle begins again.

Beekeeping Models

Beekeepers come in all shapes and sizes, ranging from hobbyists to full-time commercial operations with dozens of employees. Since you'll probably encounter all of them as you begin your foray into beekeeping, it's worth finding out where they—and you—fit in.

Hobbyists and Sideliners

The typical hobbyist has from one to five hives, and uses standard management techniques such as treatments (chemical and otherwise) and routine feeding, and for the most part, do not raise their own queens. Even with these treatments and feedings, many beekeepers lose a large percentage of their bees every year. The treatments are time consuming, expensive, decimate essential microbial cultures, contaminate equipment, and poison the bees. Note that this group includes beekeepers who use "natural treatments," and even many of those who meet organic certification requirements.

Some hobbyists opt to avoid using any type of treatments. Instead, treatment-free beekeepers are dedicated to interfering with the bees as little as possible. They take the long view, emphasizing breeding from (or sourcing) localized, untreated stock over any short-term honey harvest.

Sideliners make part of their living keeping bees. Sideliners might emphasize honey production, honey sales, honey brokering, pollination, candles, pollen sales, equipment, beekeeping lessons, and writing articles. In most cases, sideliners started out as hobbyists. Although most sideliners use treatments, a growing number are adopting treatment-free approaches.

Commercial Beekeepers

Commercial beekeepers make most of their income from beekeeping. Anyone with more than 300 hives is generally considered commercial. Some commercial beekeepers are stationary and produce honey, wax, bees, or do pollination local to where they live. Others are migratory, moving their bees to as many as 20 locations a season, often coast to coast.

The commercial beekeeping industry is responsible for pollinating the majority of the food grown in the United States. The migratory beekeeping industry could not be

Bee Aware

Two thirds of the commercial hives in the United States are trucked to California for almond pollination during the same three weeks in February. This annual pilgrimage has come to drive the rhythm and tone of commercial beekeeping, and it is not sustainable.

replaced with a stationary model, as the monocrop agriculture (where huge tracts of land are dedicated to the production of a single crop) that feeds this nation requires, but cannot support, an insect pollinator population. To change migratory beekeeping would be to change all of agriculture.

If such changes are to occur, they will happen for economic and environmental reasons, and we as a nation will require local beekeepers all over the country in order to keep our food supply safe and diverse.

Many of our current crops, such as grains, are wind pollinated and do not require bees. However, fruits, nuts, and vegetables are highly dependent on insect pollinators, especially bees. Bees will always play a valuable role in our food system.

Food pollination is one of the practical reasons for keeping bees. But for many people who keep them, practicality has nothing to do with it. Keeping bees is a lot like falling in love. Both provide a lot of sweetness tempered with an occasional sting.

Furthering Your Bee Vocabulary

Before closing this chapter and moving on to the inner workings of the hive in Chapter 2, we need to briefly introduce and explain a few more concepts:

- **Foraging** Finding and eating food. Honeybees' forage can include nectar (from which they make honey), pollen (from which they make beebread), water (which they use for everything, including cooling the hive), and plant resins (from which they make propolis).
- **Propolis** Bee product made from plant resins and used by the bees as glue, plaster, as a disinfectant, as well as to reinforce and stiffen comb.
- **Queen excluder** A mesh screen placed at the entrance of a hive box that allows workers to get through, but prevents passage by queens and drones (who have larger thoraxes). Throughout this book you will see references to both a queen excluder and a queen includer. These are precisely the same device, and what we call it depends on if it is used to keep the queen out or in.
- **Honeyflow** (or just "flow") Period in which there is abundant nectar available for forage.

- **Drawn comb** Fully formed honeycomb. Bees can draw comb from foundation (see next entry), or in an empty space.
- **Foundation** A sheet of beeswax or plastic used as the basis on which comb is drawn. It is embossed with hexagons, thereby giving the bees a "foundation" for building cells. Most beekeepers use foundation.
- **Honey Super Cell (HSC)** A plastic frame with fully constructed plastic comb. This differs from foundation in that the comb need not be drawn; it is molded out of plastic.

The Least You Need to Know

- You can keep bees almost anywhere.
- All you need to do is provide housing for the bees; they will do the rest.
- Bees' activities follow the seasons; in spring and summer they gather nectar and pollen, raise most of their brood, and store food to survive the winter.
- A small but growing number of beekeepers are refusing to use treatments on their bees.

Chapter 2

Inside the Hive: The Caste of Characters

In This Chapter

- How the hive is organized
- What workers do
- Her majesty the queen
- The essential drone
- The importance of microbes

Bees have been converting air, water, nectar, and pollen into wax and honey for 100 million years. It's an amazingly complex, yet efficient, system. Biologists apply the label *superorganism* to bee colonies, meaning that they are more than the sum of their individual components. In the case of a colony of honeybees, the organism (individual bee) is unable to survive outside of the superorganism (colony).

For all its complexity, there is no central "conductor" of any kind. No boss bee tells the other bees what to do (not even the queen). Rather, the behavior of the colony is dictated by an ever-changing combination and balance of stimuli from both inside and outside of the hive.

For example, the presence of brood (who need protein) stimulates pollen collection. The presence of stored pollen inhibits pollen collection. A balance between pollen requirements and pollen stores is reached and maintained. In other words, the bees know exactly what to do, and when to do it. Biologists call such systems complex adaptive systems (CAS).

An apt analogy would be the island of Manhattan. Everything that goes in and out of the city must travel through a set number of seaports, bridges, and tunnels.

The hotter it is outside, the better cold soda sells in the park, and the more the vendor orders. The more cold sodas sold at the park, the more frequently the recycling bins need to be emptied, and the more frequently the recycling needs to be transported off the island. Every need, want, and waste on the island of Manhattan is treated the same. Supply and demand facilitate the import and export of everything fluidly and efficiently.

There is no one in the hot seat. No one is responsible for the details of such interactions. Actually managing such an operation would be impossible. Yet the food supply is maintained, and trash doesn't pile up appreciably. The pressures, forces, and vacuums created in an open and competitive market or environment make up a CAS. There are too many variables and relationships in such a system to define things as simply "x affects y in this way." The behavior of such systems is mind-boggling.

In order to begin to understand how such a system works in a honeybee colony, we must look at its components individually.

The Workers

The bulk of the hive is made up of nonreproductive female workers. In addition to laying worker eggs, the queen also fertilizes them.

As their name implies, workers do the bulk of the work in and out of the hive, and they make up more than 80 percent of the population. Their roles are determined by their age and the needs of the colony. For instance, if all the bees assigned foraging duty (nectar, pollen, or both) disappear, younger worker bees will almost immediately take on their roles. If pollen is trapped by the beekeeper (reducing the amount of pollen brought in), the colony will assign

> **Bee Bonus**
>
> The queen takes mating flights only very early in her life. She then stores the sperm she collects from those mating flights and uses it to fertilize the eggs throughout her lifetime.

more bees to gather pollen. But keep in mind that these are not top-down decisions. Instead, they are the result of interacting and conflicting physical and chemical cues shaping the behavior of the colony one bee at a time.

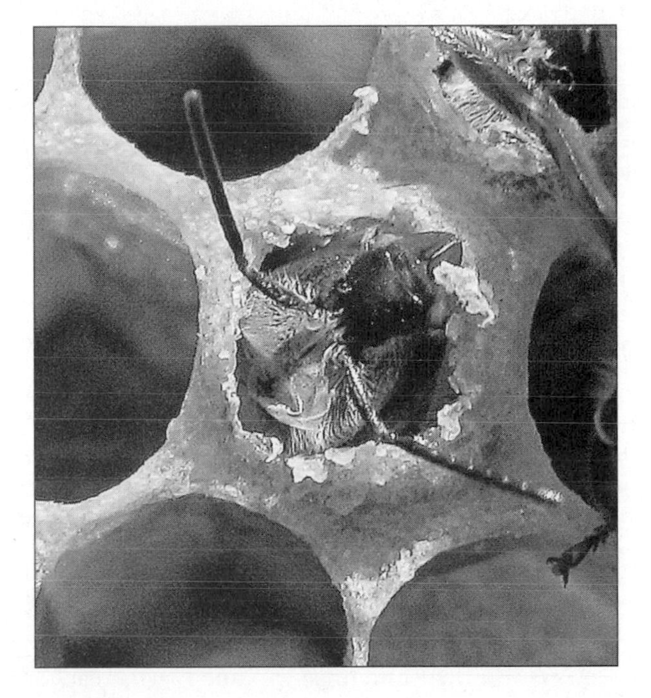

A worker bee emerging. Note the chewed wax edges of the cell and her protruding mandibles (mouth parts).

Working from Birth

When a worker first emerges from her cell, she looks a little worse for wear. Her fur is matted, her color is a bit dusty looking, and she wobbles around on the comb. As she becomes stable and takes her place in the hive order, her first task is to prepare the cells. The bees, using their tongues as their tools, must clean and polish the cells before the queen can lay in them again.

Nurse Bees

After the worker's *exoskeleton* hardens and she matures a bit, she starts to care for the brood as a nurse bee. Using her mouthparts, she feeds *royal jelly* to newly hatched larvae for three days. After that, worker larvae are fed a less potent mixture of the same ingredients (but in different proportions) known as brood food, bee milk, or worker jelly. Royal jelly is fed to queen larvae through their entire larval period. The source of protein for both of these foods is *beebread* ingested by the nurse bee and synthesized by glands in her head.

def•i•ni•tion

An **exoskeleton** is the hard shell that many invertebrates (including insects) have.

Royal jelly is a protein-rich formulation of bee milk mixed with regurgitated carbohydrates and fed to queen larvae.

Bee milk is produced from glands in the nurse bee's head (hypopharyngeal and mandibular glands specifically), and requires that the nurse bee feed on protein-rich pollen/beebread. Although royal jelly is often sold for human use, its production is extremely labor intensive, and the purity (or even authenticity) of what you can buy is questionable at best.

Beebread is pollen fermented by the bees and used as a primary protein source for developing and young adult bees.

The development of the worker larva is staggering. In the six days between the egg hatching and the bees capping it over, the larva increases in weight more than 500 times with virtually no waste!

During those same six days, worker bees visit each larva about 110,000 times for feeding and checking. The workers use their antennae to "smell" what each larva needs, and feed it with a custom-mixed formula.

Bee Bonus

Despite all the food they consume, the larva only defecates once. No one wants to poop where they sleep, so the larva holds its waste until it has finished feeding and after it has started to spin its own cocoon. Here's a little beekeeping humor for you: a honeybee larva poopates before it pupates.

Working the Construction Crew

When the young nurse bee is a little older, she graduates to comb building and capping. The bees start to fly and orient themselves as early as 4 days after emerging (while they are still cleaning cells), and by day 20 have graduated to guarding the hive entrance and foraging for nectar, pollen, and water.

The Ravages of Time and Work

In the summertime, a worker lives only about six to eight weeks. Because her body is never repaired, her wings tatter in the wind, and likely she will be outside the hive foraging when she finally has a load too heavy to lift. Very few bees die inside the hive in summer.

Cabin Fever

Winter workers are bees raised in the late fall who survive for the early pollen flows come spring. They have more fat reserves than bees raised during other times of the year and likely live longer because, for the most part, they don't expend a lot of energy raising or producing food for brood. The incredible nutrition worker bees produce for the brood takes a serious toll on their bodies.

Bee Smart

Up to 3,000 eggs can be laid in a day, and 3,000 bees can die in a day. Even if you see some dead bees in and around a hive, they are not likely to represent a significant number of those that died recently.

The Queen: The Mother of All Bees

Everyone knows that there is one special bee in the hive—the mother, the fertile daughter, even the father of all bees.

A virgin queen goes largely unnoticed, and gets no special treatment until she mates.

The laying queen, however, is the center of the hive universe. She is constantly groomed and fed by a rotating staff of *house bees*, and her scent, as it spreads through the hive from one bee to another, assures the bees that all is well. Something as simple as a damaged foot gland can alter these chemical signals and give the hive a reason to replace the queen.

def•i•ni•tion

House bees are worker bees that work inside the hive cleaning, nursing, managing stores, etc. Worker bees eventually graduate to outside work and are then called forager bees.

The queen generally goes about her business walking in spirals around the comb and laying up to 3,000 eggs a day, pausing every few minutes for feeding and grooming.

Bee Bonus

Three thousand eggs a day is about three times the body weight of the queen. The hive's investment in eggs is minimal—they do not need to be fed, for instance. Once the egg hatches, the larva needs to be fed, and the investment becomes greater and greater.

Bees are masters of economy, and will often eat worker eggs and even young larvae when food is limited and the queen has laid up more brood than the colony wants to commit to feeding. Raising brood takes time, and a small nonrefundable deposit of eggs and young brood is often a good investment. The presence of eggs and young brood means a workforce can be raised several days earlier when things turn around.

Emergency Measures

If there is no laying queen, the hive will begin the process of correcting the situation if possible. The only real difference between a queen and a worker is the nutrition provided during development. Queens and workers are fed roughly the same diet for the first three days of development, and after this time workers are shifted to a lower protein diet. Larvae that are otherwise destined to become workers can be made into queens if they are three days old or younger. If the queen is missing, the bees will immediately prepare to raise some of the worker larvae as queens.

If there is no queen for one reason or another, and it is impossible for the hive to raise a new queen—there is no brood of the proper age, for instance—the hive will do a wondrous thing. Workers (who have never mated) will begin laying unfertilized eggs that develop only into drones. This is a terrible situation for the beekeeper to be in, as drones don't appear to do anything but eat and possibly mate, and the hive will die off quickly. From an evolutionary perspective, a colony that can't possibly survive to produce new queens will use all of its energy and stores to send drones far and wide in the hope that they will mate with virgin queens, passing the genetic material of the hive to future generations.

The Making of a Monarch

The life of a queen begins much like the life of other bees, with some additional privileges and responsibilities.

A virgin queen emerges into her adult life from a queen cell. Resembling a peanut shell more than anything else, the queen cell differs from worker cells in size, shape, and orientation.

Workers completing uncapped queen cells with developing queens inside.

The new virgin is likely to first open up and sting any other queen cells in the hive, fight to the death with any other emerged virgins, and then walk around largely unnoticed by the bees for several days while she hardens her exoskeleton and builds her strength. Such a queen will mate and displace—or, in beekeeping terms, *supersede*—the existing queen sooner or later. It is not unusual for there to be a mother and daughter queen both laying in a hive, but the daughter will eventually take over.

A new queen might be raised by the hive in any of the following circumstances:

- The old queen might be perceived by the hive as a whole as unsuitable. She could be old, damaged, sick, or poorly mated. Other factors can also come into play, such as not enough bees to care for the brood, a diseased hive, or not enough food.

- The hive might be prosperous enough to reproduce itself. With lots of stores, drones, and opportunity, the population of the hive can get so great as to dilute the queen's pheromones. This shortage of pheromones, or the inability of the queen to lay because the comb is full, will trigger the swarming impulse.

  Bees sometimes issue both primary swarms and secondary swarms. The primary swarm is the first to leave the parent colony, and contains the old, mated queen. There are often several capped queen cells (and sometimes even emerged virgins) when the primary swarm leaves. Workers might drum on capped queen cells with their feet, presumably to keep some queens inside until other swarms have departed.

Bee Smart

A swarm has nothing of value to protect—no comb, no brood, no food stores—and therefore is shockingly undefensive. If you stick your hand in the middle of a swarm (yes, you can do this with a swarm, but please make sure they are not unusually defensive *first*), you feel a unique sensation of heat and vibration. It's quite an experience.

Secondary swarms are generally smaller, travel farther from the parent colony, and contain one or more of the virgin queens. Secondary swarms are often very small and might not have much of a chance to build up for winter.

- If the existing queen is disabled or missing, the bees might make an emergency queen (make a queen from larva less than three days old). The bees know within minutes if the queen is gone, and start to take measures immediately. If the queen is confined to a small section of the hive, queen cells will often be built in another part of the hive where her presence is felt less strongly.
- The beekeeper may graft a queen by removing a young worker larva from its cell and placing it in something resembling a queen cup. It is placed in a colony without a queen (along with up to 50 other similarly manipulated larvae) and raised as royalty.

  The queen cell will then either be placed in a colony where the virgin queen will emerge, mate, and become the laying queen (replacing the existing queen, or becoming the queen of an otherwise queenless colony), or she might emerge into a cage where she is held as a virgin until being placed in a hive either with or without a laying queen. If there is a laying queen when a virgin emerges or is placed in the hive, she will usually supersede her.

Bee Smart

Although some beekeepers have a prejudice against emergency queens, the truth is that given enough nurse bees and food to raise them, emergency queens are every bit as good as any other kind of queen. Good emergency queens can be reared by you, the beekeeper, from larvae that are no more than three days old. For best results, we recommend using a frame with all stages of brood when raising emergency queens.

Like all complex systems, there will be circumstances that fall outside the norm. For instance, it's not uncommon for two queens to occupy a hive. A more surprising event is *thelytoky*, where under emergency circumstances unmated worker bees lay fertile eggs that develop into properly laying queens! This unusual trait seems to be more concentrated in some bee populations than others, presumably because conditions selected for the trait, and is not something a beekeeper can rely upon.

Queens Are What They Eat

Queens are genetically identical to workers. The only difference is what they eat while developing. Extra nutrition gives the queen a larger thorax and abdomen. In addition, it enables her reproductive organs to develop completely. A worker is a stunted queen.

> **def•i•ni•tion**
>
> **Absconding** resembles swarming, except that no bees are left behind in the parent colony. The entire colony leaves its hive.

Before the queen mates, she appears very different on the comb. Her thorax is quite large, but her abdomen appears compressed and her movements are quick. After completing her mating flights, the workers feed her more and more, and her abdomen expands. A large abdomen is necessary for good egg production. The colony will thin the queen down if she is expected to fly either to swarm or to *abscond.*

Ooh, That Smell!

The queen's mandibular glands in her mouth and her tarsal glands on her feet secrete chemical cues to the workers that affect their behavior. The balance and quantity of these substances—and likely some we don't know about yet—cue the colony as to the health and well-being of the queen. The workers' constant grooming of the queen spreads these pheromones throughout the hive.

The Drones

With bees, unfertilized eggs are males (drones), and fertilized eggs are females (queens or workers).

In other words, drones have no father. *Haploid clone* is the technical term for drones, meaning that they have half the genetic material the queen has and none of the genetic material from her mates. The sperm of the drone contains all of the genetic material of the drone (again, half that of the mother queen). Because the drones have no father, when a drone mates with a queen, it is genetically two queens mating. Drones can therefore be considered "flying sperm" or "flying gametes," and the queen that produced the drone is the "father queen."

Drones don't do much active work in the hive. They don't collect nectar or pollen, and they can barely feed themselves inside the hive, preferring to be fed by the workers. Conventional beekeeping wisdom dictates that drones do nothing in the hive but use up resources. Beekeepers have even tried to eliminate drones from the colony. This is a mistake, as drones have numerous underappreciated roles in a healthy hive.

First Line of Defense

Drone brood is generally laid around the outside of the broodnest. In the case of an attack by a bear or other wild animal, the drone brood will be the first to be eaten. From the perspective of the hive, if they can drive the predator off by stinging it before it eats all the worker brood, they are ahead of the game. Placing the drones—who have no direct role in building/maintaining the colony—on the outside as "sacrificial lambs" works well as a defensive strategy.

Situating the drone brood on the outside of the broodnest defends against another enemy of the honeybee as well: cold. The brood needs to be kept warm—about 94°F—in order to develop properly. Bees keep the brood warm by covering them with their bodies, and bees keep each other warm by clustering tightly. If conditions change suddenly—a cold snap at night for instance—the bees that have been covering the brood must cluster with each other more tightly to keep the nest as a whole warm. As the cluster tightens around the hub, the bees abandon the outskirts in favor of maintaining the center. The drone brood on the outskirts is sacrificed in favor of the developing workers, who will eventually gather nectar, raise more brood, and defend the hive. In addition, the drone brood acts as a layer of insulation for the worker brood.

Hive Health

Drone brood acts as a filter, removing pests before they can infest the valuable worker brood (much as the liver removes toxins from the bloodstream). For instance, drone brood attracts *varroa mites* away from worker brood (where the functioning of the hive would be affected).

Bee Aware

A healthy colony will raise 10 to 15 percent drones. In order to do this, a colony will keep drone comb in and around the broodnest. Many beekeepers believe that drone comb should be eliminated from the hive, as drones are perceived as "lazy" and "unproductive." Actual studies by Dr. Clarence Collison, however, have not shown there to be a drop in productivity when the drone population increases, and that the bees will do their best to raise the same number of drone brood no matter what the beekeeper does. Clearly the bees know more about this than we do.

The Microbes

Microbes living in a hive perform more functions than the bees do. They hold thousands of niches in bees and in beehives. They feed on every waste, capitalize on every opportunity, and compete without mercy. Unique species of bacillus (a common genus of bacteria) live in the honey stomach of the honeybee, and contribute enzymes used to transform nectar into honey. Gut bacteria assist the bees in the digestion and assimilation of nutrients. Molds, yeasts, and bacteria take turns fermenting pollen, and it becomes beebread, which has twice the water-soluble proteins as raw pollen and is suitable for long-term storage as a "pollen pickle."

def•i•ni•tion

Microbes are life forms too small to see with the naked eye. Microbes include bacteria, yeasts, and molds.

Some microbes that can cause certain diseases in the hive are also responsible for preventing other diseases. These complex and dynamic microbial relationships will be explored further in Chapter 10.

Somehow, it all works out that none of these microbes does well in the long run if it crowds the others out. The successful microbes are those that can take advantage of a situation, but not be overly antagonistic toward the bees or the other microbes. It is a beautiful system, one in which competitive cooperation is rewarded with community and long-term survival.

Microbes that have a tendency to dominate such a competitive and cooperative environment tend to alter the environment with their waste products or by eliminating their food source to the point where their own survival becomes impossible. Such microbes are selected against in a natural system, and this is precisely why a natural system must be maintained. This microbial foundation is necessary to maintain the health and vitality of the honeybee colony.

Bee Smart

Be humble before the microbes! Be humble before such balance and intricacy! It's simply beyond our ability to master such a system, beyond us to direct or conduct such a concert. Be a joyful spectator.

The most recent assays of microbes in a beehive are now terribly out of date. Researchers identified more than 8,000 strains of microbes in beehives *before* DNA sequencing was feasible. This means that the 8,000 that have already been identified were those that could be cultured in a petri dish in a lab. Using more advanced techniques, researchers are finding more species of microbes in the hive every time they bother to look, including bacteria in the honey stomach, where none were thought to exist, making 8,000 barely the tip of the iceberg.

One must look at the microbial/bee culture closely. The relationships are not one-to-one; they all exist in a system, in a cloud of dynamic relationships that vary with conditions in and out of the hive. It's entirely possible that simply pulling frames and exposing them to the sun's ultraviolet rays causes lasting damage to such cultures. Even more damaging to the bee/microbe balance are the many treatments beekeepers use on hives in order to destroy disease-causing microbes and other pests. See Chapter 10 for details and why we believe a treatment-free approach to raising bees is best for the bees as well as the microbes.

The Least You Need to Know

- The hive is a complex adaptive system, meaning that its behavior is dictated by the interaction between its individual components and its environment.
- Workers perform tasks in a specific order, and according to need.
- The queen is mother of all in the hive.
- Drones have no father, as they are raised from unfertilized eggs.
- Microbes are the essential supporting cast of the hive.

Chapter 3

Beekeeping Essentials: Gearing Up

In This Chapter

- Rules and regulations
- Avoiding allergic reactions
- Beekeeping gear: the bare essentials
- Using a smoker
- Choosing a hive style
- Frames and foundation

In any endeavor, preparation is more than half the battle. Beekeeping is no exception. Arm yourself with knowledge and the appropriate equipment. Check on your allergy status for bee stings. Research any beekeeping regulations in your area. Get your clothing and gear together and your hive parts ready. Practice lighting your smoker and keeping it lit. Trust us, the day your bees arrive, you will be glad to have made these basic preparations.

Letter of the Law

There are fewer rules and regulations regarding the keeping of bees than one might expect. The number of hives in a residential area, their distance from property lines, and access for inspection are all regulated in some areas. Some towns and cities ban beekeeping outright, but harbor a healthy population of urban rooftop beekeepers anyway (New York City is well known for this).

Unfortunately, however, some poorly thought-out regulations are in effect, and many are on the books but never enforced. In some cases, beekeepers are required by law to register their hives, but registration forms are not available. Sometimes a call to a nearby club, association, or beekeeper can give you better information about local regulations than what is available from city hall.

Bee Aware

One law that seems to be in effect most places is that kept colonies of bees have to have inspectable comb. This means you cannot just toss the bees in a box, they need some kind of frame or top bar that will allow the comb to be removed for inspection.

We don't advocate breaking any laws, and in some cases, hive registration provides tangible protection (in our case, we get notified personally if there is going to be any mosquito spraying near our hives), but it is sometimes better to get some "real world" information from actual beekeepers before contacting state, county, or city officials.

Allergies

First of all, if you've never been stung by a honeybee, you should get an allergy test to find out whether you are allergic to them. If the test is positive, you might rethink beekeeping as a hobby, but talk to your allergist, as there are many types of allergies and ways to deal with them. Even if your tests show that you do not have a bee allergy, we encourage you to get a get a prescription for an EpiPen.

Reactions to bee stings vary. Localized reactions (swelling near the sting site) are seldom anything to worry about unless the swelling is so severe as to cause damage. Technically, any amount of swelling is an "allergic reaction," but it is not the kind of reaction that is of concern, and not what an allergy test would be looking for.

An anaphylactic reaction is another story. Telltale signs are hives or welts nowhere near the sting site, and most importantly, swelling around the eyes, nose, and throat. Such symptoms, no matter what the cause, and no matter what an allergy test indicates, require immediate medical attention, as it could be a life-threatening situation.

Sensitivities that lead to anaphylaxis can sometimes develop over time, so it is important to keep the symptoms in mind at all times, for you and anyone that is with you in the bee yard.

Bee Smart

EpiPen is a common brand name of an epinephrine auto-injector, which is used to treat an anaphylactic reaction while medical help is sought. Carrying an EpiPen is a smart thing to do even if you aren't allergic to bee stings, as new allergies can develop at any time. And keep in mind that EpiPens are designed to buy time while medical care is procured; they are not a cure for anaphylaxis.

That Lovin' Feeling

Unlike wasps, hornets, and other stinging insects, honeybees die when they sting. The bees seem to know this and are reluctant to sting in most cases. Unfortunately, the big exception is when they are defending their hive (where their offspring, food stores, and resources are located), and this is exactly when you will be interacting with them.

You will get stung. When you do, take note of the symptoms. Again, pain, itching, swelling, and redness in the area around the sting are not a concern. However, if you experience any swelling of the throat, trouble breathing, or other symptoms of anaphylaxis, get medical help immediately.

When a bee stings you, it usually leaves the stinger behind in your skin. You should remove it as soon as possible. You can use a hive tool or credit card and scrape it along the skin, pushing the stinger out. You can also grab it with your fingertips and yank it out.

Bee Smart

For normal pain and itching from bee stings, try applying a poultice of plantain leaf. This plant grows wild in most areas of the United States. Applying a poultice of crushed or chewed leaves on the sting site relieves pain and itching for many people.

A common "home remedy" is a paste of baking soda and water. Any topical antihistamine will help (especially with itching), and an oral antihistamine will help if you are uncomfortable.

Dress for Success

When dealing with hives, dress according to the strength and mood of the bees.

Wear light-colored clothing, and avoid any articles that have attracted defensive attention in your past dealings with the bees (a particular pair of shoes, a camera, socks, etc). Bees especially seem to be irritated by wool and fuzzy, textured clothing. Don't wear any artificial scents (perfume, aftershave, shampoo, deodorant, hand lotion, etc.) near the hives, as the bees don't generally appreciate such smells.

Any smell of *alarm pheromone* that bees may have left on your clothing can also be a problem. The purpose of the alarm pheromone is to alert the bees to danger. Be sure to frequently wash any clothing you wear in the bee yard to minimize unwanted attention.

def•i•ni•tion

Alarm pheromone is the scent that bees emit to alert other bees to danger. It's worth noting that alarm pheromone smells a lot like banana oil, and you should avoid eating bananas in the bee yard.

The Beekeeper's Toolbox

After flipping through a beekeeping supply catalog, you might get the impression that to be a beekeeper you need a lot of devices and gadgets. In reality you only need a few basic tools and pieces of safety equipment to get started.

Here is the basic equipment you will need.

A Veil

A veil is a most essential tool for beekeeping. Bee stings can hurt, but being stung in the eye can cause lasting damage. We don't always wear veils (especially when dealing with nucs or smaller hives whose temperament we know), but we do always have them handy. When the bees tell you (with their stingers) that you need one before you can close the hive up, you will listen.

Several styles are available but we prefer the "inspector's jacket" with a zip-on hood. Tie-down veils are time consuming to tie and untie, and they are never quite sealed, whereas the jacket just zips and can be tucked into your pants to prevent bee access. Get a jacket that is several sizes too big if possible, as bees can sting through almost any suit, and a billowy jacket that isn't touching your skin is your best friend.

Some veils have a tendency to make contact with your face (nose or otherwise) while working. This is not good, as it enables the bees to sting through what is supposed to be protective equipment. Wear a baseball hat or visor under the veil to keep the veil away from your face. Safety pin the veil to the hat or visor if it tends to slip around.

Bee Smart

The feeling of sweat dripping down probably really is sweat, as bees tend to climb upward. If, however, you feel the sweat dripping up, it is probably not sweat. When working with bees, we recommend that you tuck everything in. A shirt tucked into pants will prevent bees from climbing up under your clothing. Likewise, we always tuck our pants into our socks. If you feel something dripping up your pant leg, we recommend that you squash it through the fabric. You might get stung, but better to be stung on the leg than higher up!

A Hive Tool

Hive parts are often glued together by the bees with propolis; therefore, a hive tool is essential for prying them apart. Basically, a hive tool is a crowbar optimized for beekeeping use. Several styles are available, and it's not a bad idea to get a few styles to try out. Don't be tempted to use a paint scraper or other general-purpose tool. Good hive tools are properly hardened and tempered for the stresses they will see in the bee yard.

Gloves

We generally work without gloves, but some bees, in some circumstances, require them. We've found the gloves available from the beekeeping supply houses to be lacking. It is impossible to keep them clean of propolis and alarm pheromone (which will build up on the leather and rile the bees) without drying out the leather, and the thickness of the fingers makes handling the frames clumsy.

Instead, keep a pair of yellow rubber dishwashing gloves handy for when you need them. Bees can barely sting through them, and they are cheap, less clumsy, and widely available. Because they aren't breathable, your hands will likely sweat, so have several pairs on hand and rinse them out frequently.

A Smoker

Smoke seems to affect bees in two major ways. First, it tricks the bees into thinking there is a fire. Bees know that in order to survive a fire they must leave the hive, and their chances of survival are better if they fill up with honey before leaving. Smoking the bees makes some of the bees go head first into honey cells and ignore everything else going on.

Secondly (and more importantly), smoke has an overpowering smell, one that masks any pheromones the bees might emit. Therefore a warning or danger signal (alarm pheromone) never gets passed on to the next bee, and defensive reactions are minimized.

A smoker. The bellows move the smoke from the fire chamber up through the top opening.

You should always light your smoker before opening hives, and you should always lightly smoke hive entrances (and your hands) before opening them. Smokers also come in handy when reassembling hives, as a few puffs of smoke will help drive bees down between the frames so they don't get squashed. Smoking your clothes and any places a bee may have stung or deposited alarm pheromone helps to keep the bees from getting agitated.

How to Use Your Smoker

Lighting a smoker is an art. The smoke needs to be cool, and you have to make sure that sparks (and flames) don't shoot out and burn the bees (or the beehive, or the woods, or a house, or the beekeeper's assistant). Even if you aren't using it, keeping the smoker lit when dealing with an open hive is essential for safety, and letting it go out is grounds for a healthy ribbing by other beekeepers.

You can use whatever type of fuel is readily available in your area. In New England, the most popular type of fuel is dried pine needles. In the Arizona desert, mesquite is everywhere, so beekeepers smell like barbecue. Cardboard, leaves, or saw dust from wood that hasn't been pressure treated or glued together will do. Old twine or burlap are excellent, but it's hard to find these today without synthetic additives that can produce toxic fumes. Like any fire, you need to establish a bed of coals. Start with a piece of paper, and add a little of your fuel to get it started. Steady, full strokes of the bellows will keep things burning. Add fuel slowly until you have a bed of coals and good thick smoke.

Bee Smart

Make sure you have a good bed of coals and extra fuel on hand. If the smoke is hot, stuff some green grass or leaves on top of the fuel before closing the smoker. This acts as a spark arrestor, and cools the smoke. The purpose of the smoke is to mask other smells, not to heat the bees.

Langstroth Hives

We prefer to use what are called Langstroth-style hives. These are the stacked boxes (often painted white) that many people picture when they think of honeybees.

Bee Bonus

Lorenzo Langstroth refined the state of the art of beekeeping in the 1850s by utilizing *bee space*. Within a hive, a small space (smaller than a single bee) generally gets filled with propolis, and a large space (bigger than two bees) generally gets filled with comb. Any space, crevice, or passageway that is in this bee space range is left relatively free of obstruction by the bee.

Bee space can be understood by looking at the parallel spacing between combs. Think of it like the space (aisle) between two shelves in a supermarket, which allows shoppers moving in opposite directions with their carts to pass without colliding. Too little space causes a clog, and too much space leaves less room for products.

By designing his hive with bee space in mind, Langstroth allowed for the use of uniform interchangeable hive parts and combs that don't get glued in place and attached by the bees.

Bee space.

Langstroth hive components are available in a wide variety of sizes, most commonly in three depths: shallow, medium, and deep boxes. In addition, they come in widths of 5, 8, or 10 frame capacity. The frames are spaced such that the bees do not attach the frames to the body of the hive (proper bee space accomplishes this). Frames can be wood or plastic, with foundation or foundationless, and can even be one-piece plastic frames with plastic comb prebuilt (Honey Super Cell and Permacomb are two brands of this type).

We recommend that you decide what size equipment you will use, and use only that size. If you don't use chemical treatments, brood frames and honey frames are completely interchangeable, and a standard size will provide you with the most options for hive management.

The two main factors in deciding what size equipment to use are cost and weight. The larger the box, the fewer boxes (and fewer frames) you will need to provide the same total internal volume of space. This affects the cost of each hive (and the work of building/assembling) greatly. In other words, if you use smaller boxes, you will need more of them per hive.

Conversely, the smaller the box, the easier the box is to lift when full of honey. A 10-frame deep box full of honey weighs approximately 100 lbs., a 10-frame medium box can weigh up to 60 lbs., and an 8-frame medium box weighs less than 50 lbs.

In addition to boxes, a Langstroth hive requires some kind of stand, bottom board, and top cover(s).

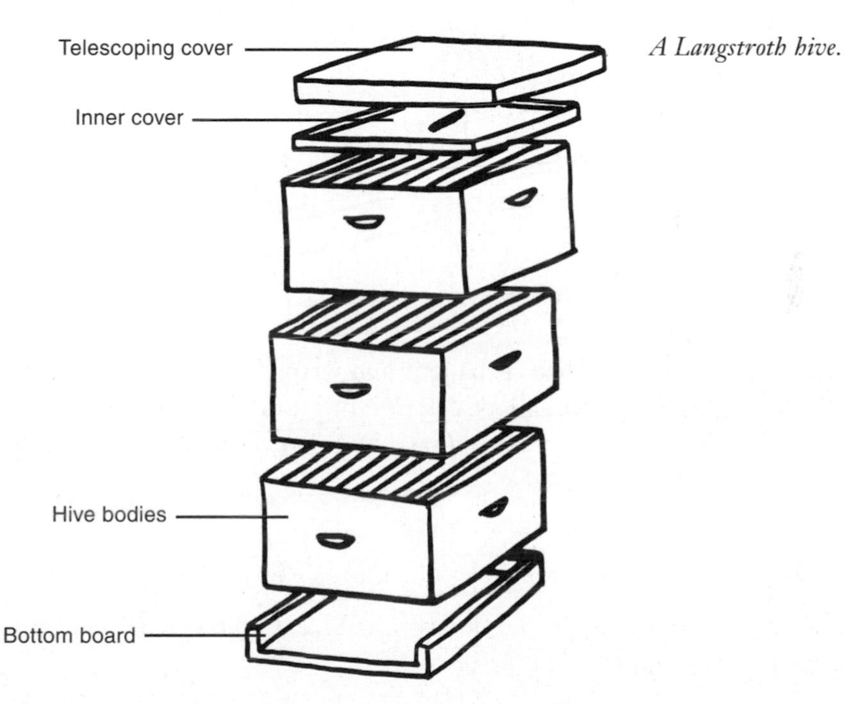

A Langstroth hive.

Stand

The stand should keep the hives off the ground at least a couple of inches. In dry areas, you might see hives with bottom boards sitting directly on the ground, but most climates require some kind of stand.

We prefer to avoid using treated lumber for hive stands, as bees sometimes drink water or spilled honey off the stand. A stand made of plain, untreated wood will eventually rot, and at least some of the time it will rot and dump a hive before you have a chance to replace it.

Bee Bonus

Amazingly, Langstroth hive components are simply stacked on top of one another, and nothing save a rock on the top cover, its own weight, and a little "bee glue" (propolis) holds it together.

Our favorite stands are made from decking (composed of sawdust and plastic) available at most home improvement stores. The decking stands on edge and is attached with high-quality polyurethane adhesive and stainless-steel screws. Last time we made them it cost about $40 a stand, and each stand holds up to four 10-frame hives. The stands should last about 30 years and never rot.

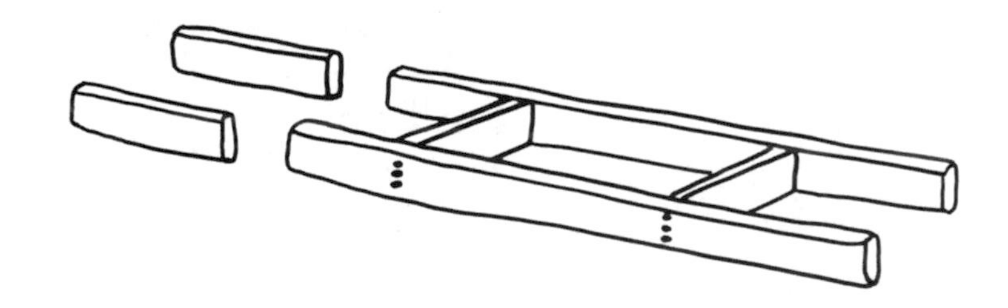

A composite decking hive stand that can hold up to four colonies.

You can use cinderblocks, bricks, pallets, and virtually anything else imaginable to make a stand. Just be careful about deteriorating wood contacting the ground, and watch for stands sinking into the soil (as a hive can weigh well over 200 lbs., 4 on a single stand is a lot of weight).

Bottom Boards

The bottom board makes up the floor of the hive. In most cases, the bottom board also provides the entrance for the bees.

For many years the standard bottom board was the reversible bottom board. This type of board can be flipped over for two different entrance sizes.

Since varroa mites came to the U.S. in the mid-1980s, screened bottom boards have become more popular. Most of them look and function much like the reversible bottom board, except that they are not reversible. They have a ⅛-inch mesh screen on the bottom, and usually a slot to slide a board underneath. Some believe that the mites fall through and can't get back in the hive; some think the increased ventilation helps; but for most, the screen offers a way to count mites as a way to measure an infestation or check the efficacy of a treatment.

Since dead bees can pile up in the winter and block the entrance, provision must be made to make sure some airflow comes in from the bottom. An *auger hole* can be drilled in a bottom box, or a small (popsicle stick-sized) wedge can be placed between two boxes. Make sure the auxiliary ventilation is on the same side of the hive as the entrance to eliminate cross drafts. If you have access to the hives in winter, a periodic sweeping out of the bottom board with a bent coat hanger wire can be helpful.

Bee Smart

A common misconception about the reversible bottom board is that the large opening is for the summer to make the entrance bigger, and the small opening is to reduce the size of the entrance for winter. In fact, the opposite is true. The small (summer) opening maintains bee space so that the bees do not build comb between the frames and the bottom board. The bees will not build comb in the winter, and the larger entrance is less prone to clog up with dead bees.

Inner Cover

An inner cover is placed between the top box of the hive and the telescoping (outer) cover, and has two functions. First of all, it prevents the bees from propolizing the telescoping cover to the top box. The trim on the telescoping cover would make it impossible to pry off from the top box without damage.

The inner cover also creates some dead air space above the colony as insulation from the cold.

Bee Aware

Insulation of the top cover seems to be essential, as the bees exhale warm, moist air when they heat the cluster in winter. If the inside of the cover is cold, water will condense and either drip on the bees, or it might freeze, and then drip on the bees when it melts.

Telescoping Cover

A telescoping cover is placed over the inner cover. Its name comes from the fact that its edges extend, or telescope, over the inner cover and over part of the top hive box.

These extended edges keep wind and rain out of the hive and generally protect the hive.

The disadvantages to telescoping covers are that they are expensive, and the overhang seems to catch the wind so that they tend to blow off of the hive (we have lost a few hives this way in winter). Beekeepers place bricks and rocks on top.

Bee Bonus

The Langstroth hive is not the only hive option, but for honey production it has no peer. Its modular design of interchangeable components allows the beekeeper to inspect the colony quickly, transfer resources between hives, make splits (increase colonies), and manipulate the bees for maximum honey production.

Leveling Your Hives

When setting up your hives, make sure they are tipped slightly forward so that any water accumulating on the bottom board will drip out rather than stagnate.

Foundationless frames especially require that the hive be perfectly level from side to side. The bees use gravity to build comb from the top bar to the bottom, and if the hive is tilted, comb may be built crossing frames, making a big mess for the beekeeper (the bees don't care).

Frames

Once you choose a hive body size, the next step is to choose and prepare the frames that will hang inside. The depth of the frames must match the depth of the hive bodies you will be using.

You can purchase preassembled or disassembled frames. Frames that come disassembled will need to be glued and nailed together with special frame nails that are available from the bee supply catalogues. (These are not standard nails, and you should not substitute standard nails when assembling frames.) Frame nails come in two sizes, and you will need to purchase both.

Because we use both foundation and foundationless frames (open frames with comb guides), we like a frame with a grooved top bar. The groove lets us insert the top edge of the wax if we're using foundation, or giant popsicle sticks as comb guides if we will be letting the bees build their own comb.

If you have a choice, order solid bottom bars as opposed to grooved bars. The groove gives wax moths and small hive beetles more places to hide.

A foundationless frame full of worker brood. Note the capped and uncapped honey in the upper corners of the frame.

Bee Bonus

If you use foundationless frames, you will get the chance to see how bees prefer to build their comb. The bees will fill some of the frames with comb so completely that they will appear to be drawn from foundation. On other frames they will leave the entire bottom edge and much of the sides of the comb unattached. This is to give the comb the vibrational potential the bees need for transmitting their dance communication. If you use foundation, you'll see that the bees will often chew out sections along the edges so that the comb gives them the proper "tuning" they need so they can feel the vibrations of the dances.

Assembling Frames

There are three distinct parts to each frame:

- **Top bar:** There are two main types of top bar—wedged and grooved. What style you use depends mostly on preference. The wedged version comes attached to the bottom of the top bar (almost cut through) for the beekeeper to remove during assembly. Note that it is a good idea to remove all wedges as a first step in frame assembly, as you don't want to nail them in place by mistake.

- **Bottom bar:** There are three basic types of bottom bar: split (with a slot all the way through the bar), grooved (the slot does not go all the way through), and solid. The choice of bottom bar is less important than the choice of the top bar. We strongly recommend that you leave an inch or two between the bottom of the foundation and the bottom bar (even if you have to trim the foundation). The grooves and slots are designed to accommodate foundation spanning the full depth of the frame.
- **Side bars:** The depth of the side bars determines the depth of the frame. Side bars also have holes drilled in order to run horizontal wires.

The frame components should be assembled with both glue and nails. Because the frames are inside the hive, we suggest using a good quality wood glue instead of synthetic glues.

Frames can be assembled one at a time with just the frame parts, the frame nails (both sizes), wood glue, and a small hammer. If all you have around is a large hammer, treat yourself to a smaller "ladies" hammer. Frame nails are skinny and don't require much force to drive them. A smaller hammer is much less taxing for this task.

1. Hold two side bars together so that they line up perfectly. Apply glue to the inside of both of the "forks" that will cradle the top bar and the bottom bar. Once you get the hang of it you can do more than two at a time, but you don't want the glue to dry before you get them all assembled.
2. Separate the two side bars, place them at each end of the top bar, and push the top bar into both side bars. Give a tap or two with the hammer to make sure they are fully seated. Drive two of the larger frame nails through the top bar directly in the center of each of the side bars (for a total of four nails). The nails should go straight into each side bar.
3. Turn the frame upside down and insert the bottom bar into each of the side bars. Make sure that the joint is tight up to the shoulder of the bottom bar. Use two of the small frame nails to secure the bottom bar to the side bars in the same way you nailed the top bar.
4. Drive a large frame nail from the side of each side bar into the top bar. If it is a wedge top bar, be careful to nail on the side without the wedge removed. This final nail is an important one, so don't forget!

You can build simple jigs that will hold 10 frames at a time for assembly, and we highly recommend using them if you have more than 30 frames to assemble. Plans are available for free online.

A pneumatic crown stapler will make the job of assembling frames much easier, and you should consider one if you plan to be in beekeeping for the long haul, and/or if you already have a compressor.

Wiring Frames

Wires help to reinforce foundation while it's being drawn (the heat and weight of the bees can stretch and warp foundation before it's fully drawn), and also reinforce the drawn comb for handling the frames and extracting.

You can purchase foundation with or without vertical wires already incorporated into it, either with hooks left exposed at the top, or without.

If you have foundation with exposed hooks, secure them under the wedge. Secure hookless foundation into a top bar groove with melted beeswax. On their own, vertical wires are not considered enough to support the comb for extraction unless handled carefully. The vertical wires keep the foundation from sagging before it's drawn out. The only real reason to use vertically wired foundation is for brood comb that will not be horizontally wired. We recommend horizontally wiring everything and buying foundation without wires.

Horizontal are installed in the frame. Pass a single wire through each of the eyelets in the frame, crossing horizontally as many times as there are holes in the end bars. The wire must be held taut and secured.

Partially drive a small nail at each end of the wire. Wrap one end around a nail several times, and then drive the nail in. Pull the wire as tight as possible, and then secure the other end in the same manner. Alternatively, you can wrap each end of the wire around the side bar and staple it in place.

It can be difficult to get the wire tight enough, and the wires also have a tendency to cut into the wood, making even tight wires loose over time. Supply catalogs sell a frame wire crimping tool that will take up quite a bit of slack without too much work.

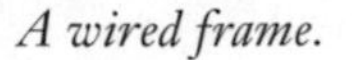
A wired frame.

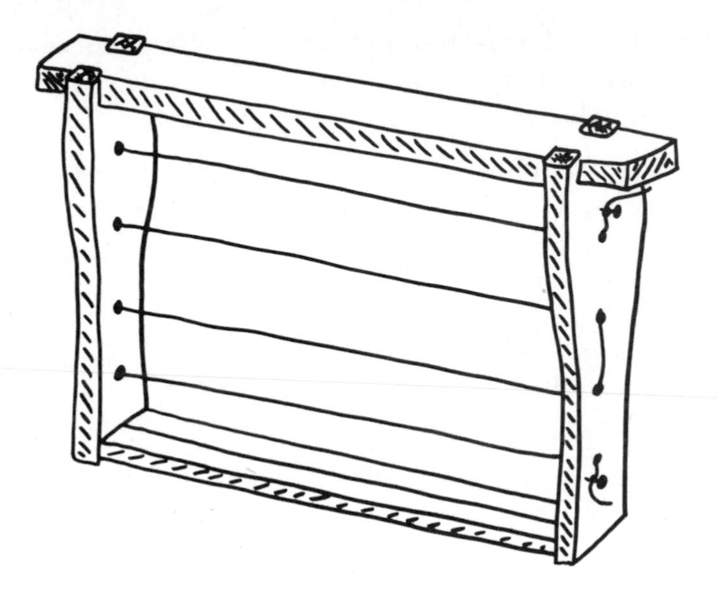

Installing Foundation

Be very careful with foundation, especially in colder temperatures, as what is flexible and pliable when warm is brittle and will break when cold. Always work with foundation in a warm room, and make sure that the foundation itself is given a chance to get up to room temperature. Also be aware that some suppliers won't ship foundation during the cold weather, so if you want to install foundation in the winter, you should order it in the early fall.

Before you begin to embed the foundation, you will need to secure the foundation to the top bar. If you have wedge top bars, hold the frame horizontally so that the side with the wedge removed faces up. Lay the foundation on top of the wires, centered side to side, with the top edge of the foundation snug against the top bar. Replace the wedge so that it clamps the foundation between it and the top bar, and nail it in place. Once you understand how it all goes together, you may prefer to partially drive the nails into the wedge before installing it.

If you are using a grooved top bar, the foundation must first be slid into the groove. Hold the frame horizontally and lay the foundation on the wires. Tilt the frame and nudge the foundation until it is in the groove for the entire length of the top bar. Turn the frame almost completely upside down so that the top edge of the foundation is square against the inside of the groove and the foundation is centered left to right. Run melted beeswax in the groove (while the frame is still upside down).

Beeswax and various tools for melting, pouring, and applying it are available from beekeeping suppliers.

Embedding Foundation

Once the foundation is mounted to the top bar, you need to press the wires completely into the wax without making holes or otherwise making a mess.

The purpose of wiring is to reinforce the foundation while the bees draw it out. To achieve the desired ability, the wires must be embedded in the foundation, and this requires some heat.

The one piece of equipment you will need no matter how you embed is a form board, which is simply a piece of wood cut almost to the inside dimensions of your frame, so that when the frame is laid over it horizontally it rests on the wires.

Beeswax is very malleable when warmed, and any of a number of means—a hair dryer, a spur embedder, or a low-wattage soldering iron, for instance—can be used to heat and embed the wires into the wax. No matter how you heat the wax, the frame should be positioned on the form board with the foundation sandwiched between the board and the wires.

The best way to embed wires into foundation is to heat just the wire by running an electric current through it. Obviously, you want to run the right amount of current through the wire so as not to burn through the foundation, or electrocute the beekeeper or the beekeeper's assistant (usually an unpaid family member).

Heat the wires until they just start to melt the wax, turn off (or disconnect) the electric current, and press the wire into the foundation with a chopstick or other embedding tool.

Foundationless Frames

Foundationless frames are nothing more than an empty frame with something running horizontally along the underside of the top bar that the bees can use as a comb guide. Actually, the bees will draw comb regardless of whether or not you give them a comb guide, but a guide encourages the bees to draw the comb where the beekeeper would prefer it to be drawn. To draw comb in an open space, the bees will cluster on the lowest point of an upper surface, drawing the comb down. It's your job as the beekeeper to give them the best guide you can.

A comb guide can consist of a strip of wax foundation (a starter strip), a wood wedge, or giant popsicle sticks (generally sold as Jumbo Craft Sticks at both Walmart and Michael's Arts and Crafts). We prefer the popsicle sticks for several reasons. The sticks can be used as is, glued directly into the frame groove, they cost about 4¢ per frame, and they are more stable and durable than wax starter strips. Some people think that the bees need the hexagon imprint on the wax to guide them to build comb, but we have seen perfectly drawn comb in hundreds of frames with wood comb guides. There is almost never a problem if your comb guides are good and you keep the frames butted up against one other in the hive.

If you prepare the guides before you assemble the frames, you can lay the top bars in rows on a table and do your gluing all at once. Apply dabs of a standard wood glue in the grooves (don't overdo the glue—the pressure of inserting the sticks will help the glue to spread) and insert the sticks along their long edges. Each frame will need about two and two thirds sticks. Break the third stick by hand or cut it with a utility knife. Don't worry if the sticks splinter at the break—the bees don't care. A gap of up to an inch between sticks or at the sides is not a problem—the bees will fill it in with wax.

Foundationless frames require careful handling, especially when newly drawn and/ or full of honey (which is heavy). Balance the ears on the top bar on your fingers, and always keep the comb perpendicular to the ground to avoid comb failure.

Foundation vs. Foundationless Frames

Since foundation came into widespread use in the late 1800s, its adoption has become nearly universal. Foundation virtually eliminates drone comb and provides very straight, even, and interchangeable combs that are structurally sound even while under construction.

Foundation has its drawbacks. It is expensive and fragile, has been shown to contain pesticides, imposes an unnatural hive structure (all worker comb), and it actually seems to take longer for bees to draw out foundation than it does for them to draw comb with only a comb guide.

Most beekeepers don't even know that bees can be kept without foundation. A common story we hear from beekeepers is that they put an empty frame in the hive, and the bees built only drone comb. They didn't want the drone comb, so they removed the frame and replaced it with another frame, which the bees again drew full of drone comb. Their conclusion is that bees will only draw drone comb in a foundationless frame. What the beekeeper is missing is that the rest of the broodnest is near 100 percent worker comb, and the bees' natural tendency is for 10 to 15 percent of their comb to be drone cells. If the drone comb is left in the hive (moved off to the side), the bees will start drawing worker comb when they have enough drones.

Top Bar Hive

The other major style of hive is the top bar hive. As the name implies, it has no frames, just top bars from which the bees draw comb. Most top bar hives are horizontal and do not have stacking components. The bars usually butt against one another along their entire length, and the bees have no access above the bars.

Horizontal top bar systems offer the following advantages over Langstroth boxes:

- No lifting of boxes
- Bees are easy to work with, as guard bees are not necessarily alerted that the beekeeper is present
- Wax is routinely harvested along with honey
- They are inexpensive to put together

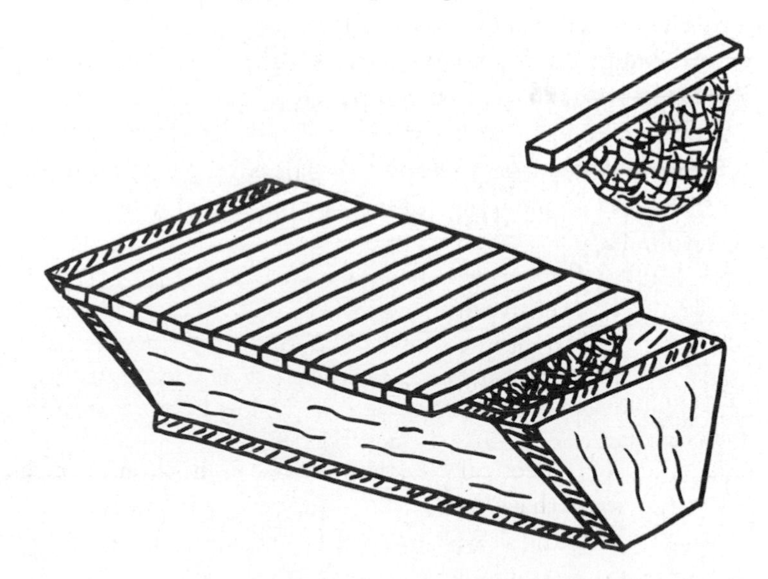

A simple top bar hive.

The major disadvantage of top bar systems is that you simply can't get the same level of honey production that you can in a Langstroth hive. The biggest mistake that one can make when designing a top bar hive is making it too small. Top bar hive dimensions vary, but think of volume in terms of three deep Langstroth boxes (about 7,800 cubic inches). You will learn a lot about bees with a smaller configuration, but you will not have the stability or production of a larger population.

Top bar hives may be vertical as well. One style, the Warre hive, has a devout following worldwide. In oversimplified terms, it is somewhat like a Langstroth hive without frames (only top bars), and management is limited to adding boxes to the bottom, and harvesting comb and wax a full box at a time from the top.

There are a variety of sources of plans, hives, and top bar–specific information available online.

A quick note should be made regarding *skeps*. These iconic inverted straw basket hives are still in use in some specialized beekeeping operations, such as heather honey from the British moors, but because the comb cannot be removed for inspection, they are illegal in most states in the United States. Their management is generally limited to collecting swarms and killing bees to harvest the honey.

To Paint or Not to Paint?

White or light-colored painted hives are the standard for most beekeepers. Well-painted and maintained hives will probably outlast unpainted hives. However, it is debatable whether painted equipment is actually better for the bees. We painted our early hives but for the past few years have used completely unpainted equipment. The wood isn't looking any worse for wear and we have saved ourselves hours of work. The original bright pine fades after a season to a pale silvery gray reminiscent of Cape Cod shingles. Instead of announcing "bees!" our unpainted hives blend discretely into the background.

Bee Smart

If you choose to paint, use nontoxic paint and give it plenty of time to off-gas before you install your bees. Never paint the insides of the hive, bottom board, frames, or any other part of the hive that will contact the bees' living space.

Mistakes to Avoid

As you begin your foray into beekeeping, you will make mistakes, probably lots of them. Fortunately, bees come with excellent "reminders" (also known as stingers) that will correct many of them for you. As for the rest, we encourage you to do your best to learn from our mistakes.

Most of this information will make more sense after you actually get some experience in the hive. However, it is never too soon to start thinking about these concepts. Bookmark this section and memorize it as best you can—you'll be returning to it again and again.

Preparedness Mistakes

Plan ahead. Don't get to the bee yard and realize you forgot your smoker, veil, hive tool, or most frustrating of all, matches or a lighter. Keep some basics in your vehicle: a hive tool, smoker, matches, and even an extra folded veil. An empty cardboard box is useful should you need to grab a swarm. Think ahead as to what equipment you might need to expand hives, split them, move them, or harvest honey, and bring it with you to the yard. Light your smoker and secure your veil before you start to inspect. Most of all, be calm and gentle, and have a goal in mind. Don't meander in the hive; work purposefully, and don't leave it open longer than necessary.

Conduct Mistakes

Keep a presence of mind about you when you are in the bee yard. Don't stand directly in front of a hive entrance. Avoid casting shadows over the hive when the top is removed. Move carefully and purposefully. If you get stung, remove the stinger and smoke the sting to prevent more stings. Listen to the hive and puff a little smoke if they start to get restless. Be mindful of where you put boxes and frames. Be mentally prepared to be stung, and stay calm.

Manipulation Mistakes

Do less rather than more. Don't move or reorder frames without a purpose. Don't break up the broodnest. Always pull frames from the least populated portion of the hive first. Always consider that the frame you are pulling or handling could be hiding the queen. Keep frames butted up against one another.

Management Mistakes

Follow the bees' natural rhythms. Don't try to expand them too quickly. Don't try to split hives before they have reached critical mass. Unless you have a really good reason, don't open hives more than once a week.

On occasion you will lose a hive. It is always a sad event, but don't be distressed. Culling is what nature does best, and as beekeepers building up a population of acclimated bees, culling is what we need the most.

The Least You Need to Know

- Find out if you are allergic to bee stings *before* you get your bees.
- A veil, hive tool, smoker, and gloves are essential beekeeping gear.
- Always keep your smoker lit and ready for use when dealing with an open hive.
- Langstroth and top bar hives are the two basic types of hive in use today.

Part 2

A Bee of One's Own

Should you buy bees or try to find them for free? Maybe a swarm settling in your eaves is what prompted you to buy this book, or perhaps you have a friend who keeps bees and can spare enough resources to get you started.

In the following chapters you'll learn about installing your bees in their new home, watching them build strength under your care, and even establishing new colonies using only your own resources.

Performing a hive inspection is more than just ripping boxes apart, and you should always have some purpose in mind when disturbing the colony. Fortunately, sitting outside and observing the bees' comings and goings can tell you a lot about the state of the union inside.

Chapter 4

Acquiring Bees

In This Chapter

- Where to get bees
- Free bees versus purchased bees
- Trapping swarms
- Transferring cutouts
- Making your own splits

When we bring an observation hive to the farmers' markets, we get a lot of questions from kids. One of the most common is, "How did you catch those bees?" It's a remarkably astute question, and one that adults rarely ask—they seem to be more concerned about how the bees find their way back to the hive when they forage. For a beginning beekeeper, acquiring healthy bees from a reliable source is your most important job.

We purchase some of our bees. We also breed our own bees, capture some swarms, and remove established colonies from walls. The best option for obtaining bees really depends on your goals and circumstance.

Where to Get Your Bees

The best source of bees is always from untreated *locally adapted stock*, be they *feral bees* or from a local beekeeper who breeds his or her bees and doesn't treat them with chemicals or treatments of any sort. If you plan to use small cell comb (which we strongly recommend; see Chapter 9), you should give some consideration to the size of the bees you are getting. If you are trying to get into beekeeping with as little capital as possible and you have the time, swarms and cutouts are your best bets.

def•i•ni•tion

Locally adapted stock is made of bees that can generally survive in your area without treatments. This is achieved through the breeding of bees that survive (and thrive) without treatments and not breeding from bees that die without treatments.

Feral bees are unmanaged honeybees living in the wild.

Buying Bees

Things can go south very fast when trying to capture your own bees from the wild. There you are, up on a ladder, nose-to-nose with thousands of bees, trying to get them to follow you home. It's a pursuit better undertaken once you have a bit of experience under your belt. So for now, buy some bees. Bees generally cost somewhere between $65 for the least expensive package, and $250 for the most expensive nucleus colony.

Bee Bonus

Local untreated bees are hard to come by. They also represent a huge niche that needs to be filled, one that hobbyists and sideline beekeepers can step right into.

One of the things you pay for when you buy bees is predictability. Predictability doesn't necessarily come cheap, but as a beginning beekeeper, it is worth the extra cost.

Nucs

A nucleus hive (or nuc) is a hive in miniature. It is a fully functioning colony of bees. Made up with frames from a larger colony with resources to spare, it is given a queen, a queen cell, or a frame of eggs and brood from which to raise a queen.

When deciding whether to buy a nuc, you need to ask a number of questions:

Where did the queen come from? If an introduced queen (or queen cell) is locally produced from stock that is treated as little as possible, then you have a nearly ideal situation.

How much honey does the nuc have? Any nuc should have stores of capped honey, beebread, and brood in all stages. The less forage that's likely to be available when you buy the nuc, the more stores the nuc should have.

How far will the nuc have to travel? If the nuc is to travel a long distance, make sure there aren't too many capped brood ready to emerge. Emerging brood can quickly overpopulate a small nuc in transit, making cooling the confined cluster difficult or impossible.

Generally, nucs are sold in sizes ranging from three to five medium or deep frames. Place the nuc into an 8- or 10-frame box of the same depth as the nuc frame (a medium box for medium frames, a deep box for deep frames). If the nuc isn't overflowing with bees and stores, you can open it in place for a few days before swapping out the small nuc box for a larger hive body. By this point, the bees will have oriented to the location of the nuc box and will enter any box placed there.

Although nucs exhibit similar behavior to full grown colonies, several differences are worth taking into account.

A nuc does not have the critical mass of bees that a full-sized hive has. It does not have the field force of a large hive and can't take full advantage of a nectar flow or pollen availability. It does not support as many guard bees, nor are the guard bees as expendable (to die in the act of stinging) as they are in a full-sized hive, making the nuc much more vulnerable. Nucs cannot quite double in size every three weeks like a larger colony can—there is too much overhead for too few bees. A hive needs to build up to six to eight frames of brood to do this.

Nucs tend to see themselves as needing more workers, presumably to be able to gather more forage and to grow the colony to the point where it can reproduce or at least survive a winter and/or *dearth*. This makes nucs an ideal place for good worker comb to be drawn.

def•i•ni•tion

Dearth is a period of time when little forage is available for bees. It is the opposite of a flow.

Nucs grow more quickly than packages (discussed later in this chapter), and are superior in many ways. However, they are not always available as early as packages are. In addition, if the source uses large cell comb, you will have to rotate it out if you

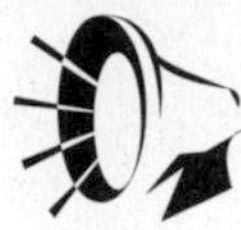

Bee Aware

Because they do not have a critical mass, nucs can go from "doing fine" to "overcrowded and swarming" or "robbed out and starving" in a very short period of time. Keep a close eye on them.

want to regress to small cell. And if the source treats, you have to get the contamination out of your equipment, which probably means culling some of your first drawn wax as well as the comb you just paid for.

And keep in mind that a nuc made up of locally adapted worker bees and an introduced mated queen from far away (or from stock that is heavily treated) offers no advantage. You are much better off finding a locally produced quality queen no matter where your bees come from.

Queenless Splits

A queenless split is a nuc that raises its own queen. If there are localized bees in your area, the new queen might mate with them (which is a good thing), or she might get eaten by a bird on her orientation or mating flights (which is a bad thing from your and the queen's perspective, but pretty good if you are the bird. It's probably good for the bees long term, as it selects for faster, more agile queens).

If you can find a local beekeeper who raises his or her own stock and uses minimal treatments, ask if you can buy five frames of bees with young brood, eggs, and food stores. The bees will raise their own queen from the brood or eggs and she will mate locally.

A month after the split, you should find brood and a laying queen. If not, pop in another frame of brood and try again, or combine the nuc with another colony.

Bee Smart

If a split looks weak and the cause is not acute disease, try giving them a frame of capped or emerging brood from a strong colony to give them a boost.

In most cases, you should only make a queenless split from a parent hive that has eight frames or more of brood and when you expect forage to continue for a while. Splitting off two or three frames of brood from such a hive will impact the parent hive, but if there is forage available it will bounce back quickly, and soon you will have two strong hives instead of one.

An ideal split is a full deep box split from a parent colony that is at least three deep boxes tall. If young brood occupy both parts of the split, and both the split and parent hive are strong, it isn't important where the queen is. All you need to do is find a good box with brood and nurse bees and put it between two boxes of comb and/or foundation, or on top of two boxes of foundationless frames (so the bees can work down from the cluster). Make sure that both colonies end up with eggs and open brood.

If the bees have to raise a queen, it will be a month or so before there is a laying queen, and another three weeks before the first new workers are born, for a total of about seven weeks. During this time the population of a queenless split cannot increase beyond the hatching out of existing brood, which will be finished three weeks into the seven-week period. Don't expect a queenless split to increase in population until these seven weeks are up. It will be two months before things really get going.

Take note that if the *field force* is out foraging for food when you remove a hive, they will return to the original location. If you are in a situation where you want to make a split with fewer bees, you can move the parent hive and leave one or two frames of nurse bees, brood, and eggs in the original location. The foragers will take on some of the nurse/house bee duties when they return to the hive, as well as provide ample forage to grow the hive quickly.

def•i•ni•tion

A **field force** consists of bees in a colony that forage for nectar, water, propolis, and pollen.

Package Bees

Package bees are an odd product. Producers shake bees out of a number of hives into 3 and 4 lb. boxes, add a can of sugar syrup, usually a dose or two of medicine, and a caged queen who is not related to the shaken bees.

Bee Bonus

The mated queen that is introduced to the package probably has nothing genetically in common with the bees that make up the package. If you've ordered a dark "Russian" package, and just see bright yellow bees, wait three to four weeks and see what the emerging brood look like before you judge. Remember that the queen has been mated before being introduced to the package in a cage.

If they could, the bees would kill the queen by smothering her (because they are unfamiliar with her), essentially cooking her to death. Since she is in a small cage at the middle of the cluster, the workers in the package can't quite reach her to do this. Logic would dictate that their second instinct would be to starve the unfamiliar queen or kill her through any means necessary. But bees don't study logic, and their second instinct is to feed and care for this strange queen through the screen that keeps them from killing her. Eventually, through prolonged contact, the workers recognize the queen as their own.

Package bees were developed as a way to ship bees without comb, and packages were regularly and successfully shipped and insured by the Post Office all over the United States for many years. It was common for beekeepers to get a phone call from their local Post Office at 5 A.M.—"Your bees are here. Can you please come and get them immediately?"

In recent years, the Post Office has stopped insuring package bees to many zones, and some of the other shipping companies have restrictions on live cargo as well. It's not as easy to get package bees as it once was, especially if you live in an isolated area.

Packages are rarely made in the north. The longer seasons in Georgia, Florida, and the west coast are more conducive to package production. The bees are fed heavily and encouraged to raise huge numbers of brood early in the season.

Bee Aware

Most package bees are from heavily treated stock and are themselves heavily treated before shipping.

Needless to say, packages are not a natural situation for the bees, and bees are not adapted to endure this kind of handling. It is not uncommon for a hive to supersede the package queen soon after you introduce the package to the hive.

Not So Local

We've been talking a lot about the value of local stock. Keep in mind that buying bees from local suppliers does not ensure—or even imply—that they are of local stock. They may simply be imported by the local beekeeper. If you are getting packages and nucs at the same time they become available from the major suppliers (i.e., late March and early April), they are most likely not any more local than what's coming from the major suppliers. By no means does this mean you shouldn't buy such bees; in many cases it's your best option.

If you live a long way from major producers, it is much better to have the bees—either nucs or packages—driven long distances rather than having them shipped. Shipping companies generally do not employ beekeepers who would be aware of the special handling that the packages need. Most bee clubs have at least one person who drives packages long distances for other members. If you are buying enough, it may pay to do the driving yourself. The less time the bees are confined the better. The cooler the bees are kept during transport, the better. How the bees should be transported depends greatly on how they are prepared and packaged. The supplier is the best source of advice on how to transport their product.

Bee Aware

If you decide to transport bees yourself, make sure to plan your trip for the fewest stops possible. Because bees can overheat very quickly, they should be shaded, out of direct sunlight, and the area of the vehicle where the bees are located should have good air circulation. If you must stop, be brief and be sure to leave the air conditioner on or the windows down. *Always provide more ventilation than you think you will need.* It's easier to make the trip nonstop if you travel with a friend and share the driving.

For detailed instructions on how to transfer your purchased package or nuc into your own hives see Chapter 5.

Free Bees

Some of the funniest stories in beekeeping are about new beekeepers with very little experience trying to catch their own bees. Of course, most of these incidents were not humorous at the time. Retrieving bees can be dangerous, especially if climbing is required to reach a hive. And because swarms often gather on high spots, a ladder is almost always required—most often when one is not available!

If you have no experience with bees, we can't recommend going after free bees. There are too many unknowns to be able to reduce the process into a simple set of instructions. All the same, we know that the kind of person who would be well suited to go after bees with no experience would never heed our warnings. So happy hunting, and please feel free to share your stories with us—we love a good laugh!

Bee Smart

If you can tag along with an experienced beekeeper to get your "bee legs" either in their apiary or on call to capture bees, do so. Once you are used to handling bees in a hive, swarms are easy to deal with.

The information offered here should set you in the right direction, and you will know much more the second time around than you did the first.

Swarms

If a swarm issues from a feral colony in your area, it is reasonable to assume that it has desirable genetics. At the very least it has some traits that allowed its ancestors to thrive long enough to reproduce, and that's a good start. If the swarm is from a treated colony (either from another local beekeeper or migratory bees that have come through your area), you are essentially getting propped-up stock that is less attractive, and less likely to be well adapted to your area. If the queen in the swarm is marked with a spot of paint on her thorax, it came directly from a beekeeper.

Bee Bonus

Before leaving the parent hive, workers preparing to swarm fill their honey stomachs to the point where they have difficulty bending their bodies around to sting, making them relatively harmless.

It is a profound experience to be present when a swarm leaves the parent colony. The bees generally move as a swirling cloud and eventually settle on a branch, fence, building, or some other structure. Here, they cluster into an undulating ball.

A swarm from one of Dee Lusby's untreated small cell hives in Arizona.

Because the bees in a swarm have no home, no comb, no brood, no food stores, nothing to protect (the bees are attracted to the queen, but they are not protective of her), they are very easy to deal with and to collect. You simply brush or knock the bees into a box or a bucket, and pour them into a waiting hive.

> **Bee Aware**
>
> As stings on the face are painful and can be dangerous even if one is not allergic to bee stings, it's always a good idea to wear a veil when capturing a swarm. We don't wear gloves when catching swarms and never get more than a sting or two, usually from bees falling directly on our hands.

When a swarm is knocked or firmly brushed, the mass of bees fall. The farther the bees fall, the more of them will take flight. Therefore, you want to have your receptacle as close to the dislodged swarm cluster as possible. A very light misting of water will keep some of the bees from flying too much. If you are brushing, you want to push the whole mass off of the object they are clustered on.

It's difficult to describe what we mean by knocking a swarm. Sometimes, if they are on a thin branch, you can snap the branch quickly up and down (as if you were shaking off water) with your container underneath. Sometimes you can get a box or bucket around the swarm, and knock it from the side with the container itself. The key is to get the greater part of the mass of the swarm to fall into your container. If the queen settles in your container, the rest of the bees will follow. If you don't have an empty hive or nuc box handy, just about any box will do for temporary use. We've used standard bankers' boxes and cardboard shipping boxes. No matter what you use to put the swarm in, the bees will need something to cluster around for support. A sturdy branch that rests on the bottom and side of the box will do just fine.

> **Bee Aware**
>
> As a safety precaution, if you are capturing bees from a populated area use queen mandibular pheromone (available from a beekeeping supply house) to attract any stragglers that may be left behind after you are gone. The bees will cluster around the pheromone and you can return the next day to retrieve them. This will help you avoid the unfortunate situation of having anyone bothered by angry, abandoned bees.

If at all possible, leave the box with the bees in place near the capture location until dark, when most of the bees return to the colony. Make sure you leave an entrance for the bees to get into the box. After dark, secure the box and bring them home.

Make sure they can breathe and circulate air, and if they are in the car with you, make sure they cannot get out, and be mentally prepared in case they do. After you get the swarm home, leave it overnight in the basement or some other cool place.

If you already have bees, you can take good advantage of honeybee behavior when capturing the swarm. The smell of open (uncapped) brood is attractive to bees, probably for similar reasons that we find babies and puppies endearing and deserving of our concern and attention. If the location of the swarm is appropriate, you might be able to place a frame of open brood against the cluster, either by tying the frame to a pole and pushing it up, or hoisting it on a rope that is slung over a higher branch. The bees will move to surround the frame, which you can then carefully lower and place in a hive box.

Bee Smart

Using drawn comb is great for splits and growing hives from nucs, but putting a swarm in a box full of drawn comb is wasting an opportunity to get fine new comb drawn.

Under natural circumstances a swarm in a new cavity needs to make comb for brood and food before colony growth can commence. Building comb is what swarms are prepared to do best.

You will want to hive the swarm quickly so they can begin to build their colony. Dump the swarm into the hive cavity, or place a bed sheet on the ground in front of the hive with one in the entrance. When you dump the bees onto the sheet, they will march up and into the hive.

Sometimes hived swarms are not immediately pleased with their new home, and can abscond. However, a newly hived swarm is much less likely to abscond if the queen can't leave. To prevent the queen from leaving, all you need to do is add a queen includer, which is a screen wide enough for workers to pass through, but narrow enough that the queen cannot. Once there is brood in the hive, the bees will stay, and you can remove the includer.

Cutouts

A cutout is a swarm that has settled in an unwanted location, such as someone's house, shed, or garage, and has made more or less progress toward building comb, rearing brood, and producing honey and beebread. Cutting into someone else's walls to remove bees requires confidence (and perhaps stupidity), even when you're doing so at the owner's request.

Much like the difference between a package and a nuc, a cutout has a head start over a swarm in that there is already comb present with brood in all stages of development as well as stores of honey and pollen.

The general procedure for taking a cutout is to locate where the bees are established and remove a portion of either the inside or outside wall of the building, exposing the bees. Using a knife, carefully cut the brood combs from inside the structure and transfer them, with all adhering bees, into frames. At some point in this process you might be able to spot the queen; make sure she ends up in the box. Then fill the cavity to prevent foragers from returning and to prevent another swarm from moving in.

Leave the cutout hive, with its entrance open, or a separate bait hive with open brood, as close as possible to the original location. After a couple of days, close the hive up and move it sometime during the night when all the bees are home.

Bee Smart

When doing a cutout from a house or other structure on private property, be sure to find out from the property owner where the bees are located. In all cases, agree on your terms in advance and put them in writing. Beekeepers don't generally do reconstruction (repairing a wall that had to be cut into to remove the bees), but the beekeeper should be prepared to return to pick up the bait hive at night or very early in the morning, before the bees start to fly and to be able to respond to any follow-up issues.

Homeless bees left behind are likely to be very cranky and will sting liberally. This is not what you want, and it's not what the property owner was expecting when they called you. It's easy enough to return at night and get them all. The few remaining bees can be drowned with a spray bottle of slightly soapy water. Remember that bees apart from a hive will die, as will any bee that stings. In situations where bees are left flying and a bait hive cannot be left, an empty box with commercially available queen pheromone will attract the stragglers. You can then collect this box at night in the same way you would collect a bait hive.

Keep in mind that it's important for the homeowner to be happy with the experience so they will call a beekeeper in the future, and not a pesticide applicator. When dealing with the public, you represent all beekeepers; do so in a way that makes us all proud to call ourselves beekeepers.

Swarm Trapping

In most parts of the country, it's fairly common for feral colonies to swarm. But chasing them around requires transportation and time. That's why it's never a bad idea to set up swarm traps and let the bees come to you.

A commercially available *pheromone lure* is one of the best ways to trap a swarm, but it is expensive. If you raise queens or requeen from time to time, you can make your own lure by storing the *pinched queens* in a jar of alcohol. A dab of this tincture inside the bait hive or on a popsicle stick placed within the hive works well. This tincture is not the same as a commercial lure but it seems to work and is essentially free. Short of actual pheromones, essential oil of lemongrass or lemon extract (typically used for baking) are reputed to work well as a lure.

def•i•ni•tion

A **pheromone lure** is bait made from worker pheromones used to attract swarms.

Requeening is the act of removing a queen and either replacing her with a new queen or letting the bees make a new queen.

A **pinched queen** is a queen that has been killed by the beekeeper by squeezing her between the thumb and forefinger.

Traps can take many forms, and it really depends on what you have on hand. An old 5- or 10-frame box with a few old frames works well, as this smells like a beehive to the swarm scouts, whose job it is to find the new home. A swarm likes to cluster in its new home. If you're using a framed trap, insert a few foundationless frames in the center so the bees have room to cluster. The trap doesn't have to have a proper top and bottom; scrap plywood will do. You do, however, want to make sure that not too much water can get in when it rains. Be sure to crack the top so that the bees have a space to get in.

Another style of trap we've seen is made from large fiber flower pots. With one slightly smaller than the other, they are stacked and screwed together, mouth to mouth (the top one upside-down). The resulting container is suspended from the top, with a hole in the bottom. Bait these with whatever swarm lure you can manage, and watch to see if bees are flying in and out. Once inhabited, you can secure a breathable sack around the bait hive from the bottom and tie it around the top to transport them back home. This is best done at night, when the bees are all inside the trap. The bees can survive a few days if kept in the shade and you keep the sack hosed down to give them moisture and to keep them cool.

Bee Smart

When you plan your swarm trap entrances, make sure you think ahead. If a swarm occupies your trap, how are you going to close things up so you can transport it? If you don't use frames, be prepared to cut out the colony that moves in.

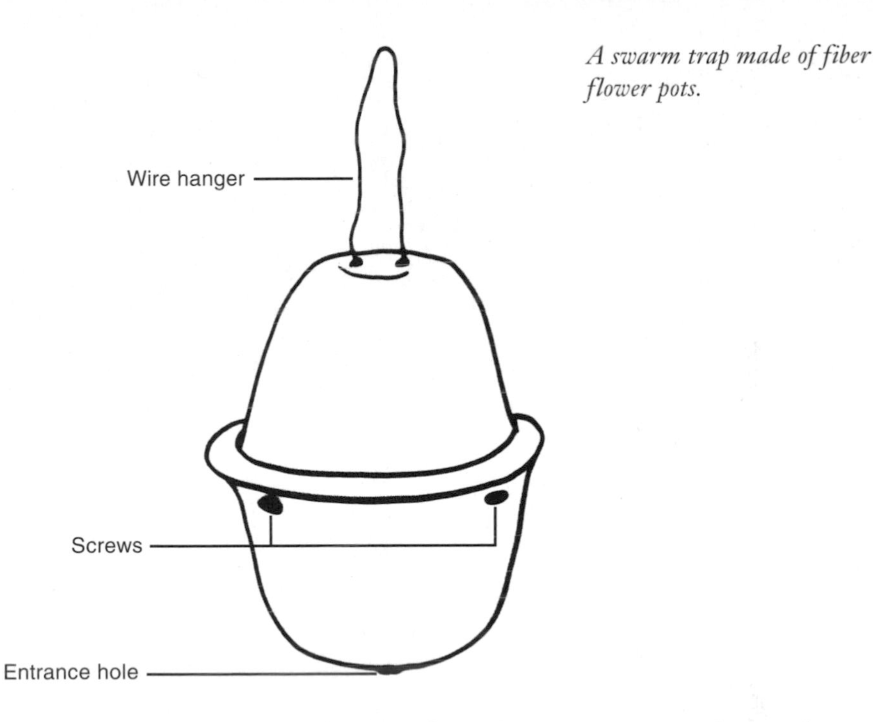

A swarm trap made of fiber flower pots.

Ideally, you want to place a swarm trap high up, but a low swarm trap is better than no swarm trap. If possible, place it somewhere that it can be monitored, so people can tell you if bees are flying into and out of the trap.

A few pieces of spare equipment in the bee yard do double duty as both onsite spares for splits and growing hives and as swarm traps. These need not be complete hives. Using the top cover of one hive as the bottom of a swarm trap is an efficient use of equipment in the field. Just make sure the box smells like bees, and the inside is sheltered. A little lure can't hurt.

Transferring Comb to the Hive

Whether you do a cutout or a frameless swarm trap, you are going to want to tie the newly formed brood comb into your hive frames.

If your cutout comb is in several pieces, "swarm ketching frames" (as made by beekeeper Dee Lusby) are very effective. A pair of simple frames is made that are exactly half the width of a standard frame. You wire these half frames horizontally between the centerlines of the side bars and crimp the wires. You then piece together the cutout brood comb with the wires.

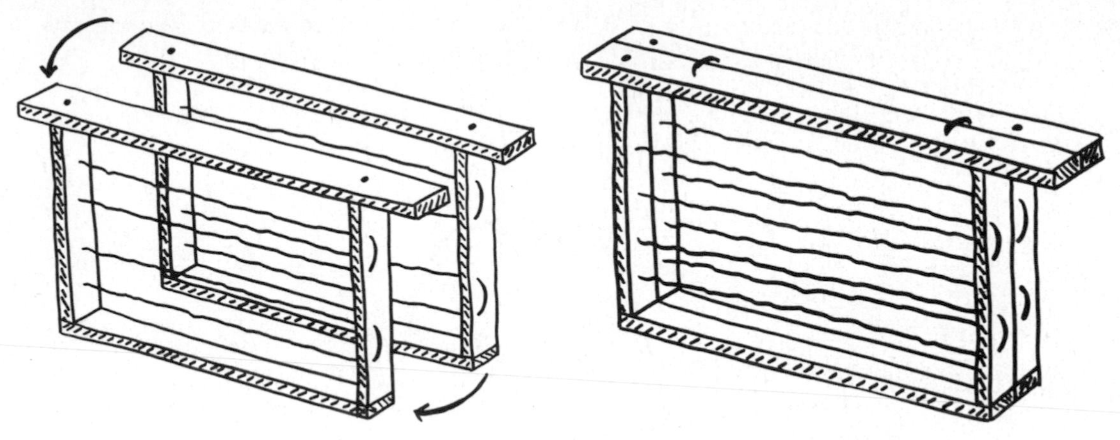

Dee Lusby's "swarm ketching frame."

The chief advantage of the swarm ketching frame is that it enables you to piece together large and small sections of comb. The bees will fill in the gaps, and you can extract honey from them.

Bee Aware

Don't try to tie in comb filled with honey. You will make a mess and likely drown bees. Put honeycomb in a covered bucket and feed it to the bees if you think they should have it or save it for later.

Lay the sections of comb horizontally onto the wires on one half of the frame. When the frame is as full as you are going to get it, place another half frame over the first one, and staple the halves together. Some brood will be damaged by the wires, but the damage will be negligible when you have the nurse bees and the rest of the brood.

If you have a single large piece of cutout comb, the Heilman-style frame is a better solution. It requires no special equipment (besides nails, string, and rubber bands), and the frames remain spaced normally (they are completely interchangeable with your other frames). This works well for unwired comb that falls out of its frame, or comb from a cutout that is trimmed to fit within the span of a frame.

Hammer a series of frame nails into the top and bottom bars, leaving about ⅛" of nail length exposed. Because all the nails go between the frame spacers on the top bar, the nails do not interfere with normal spacing or functioning of the frame.

Tie a rubber band to a piece of string (kite string is about right), loop the rubber band over the nail furthest to one side of the frame, and weave the string between the top and bottom bars until you reach the nail at the far end of the frame. Cut the string, and tie this end to another rubber band so that there is tension holding the string taut. Repeat on the other side of the frame, and you are done.

When using this type of frame, lay the frame horizontally and undo the top string. Lay the comb into the frame, resting on the bottom string, and then restring the top. These strings don't cut too deeply into the comb (they are half a frame's width further apart than the wires in the swarm ketching frame), and they don't hold small pieces of comb together. For large pieces of brood comb, however, these work splendidly. The bees will attach the combs to the frames in a short time, and you can either remove the strings and rubber bands or let the bees do it for you.

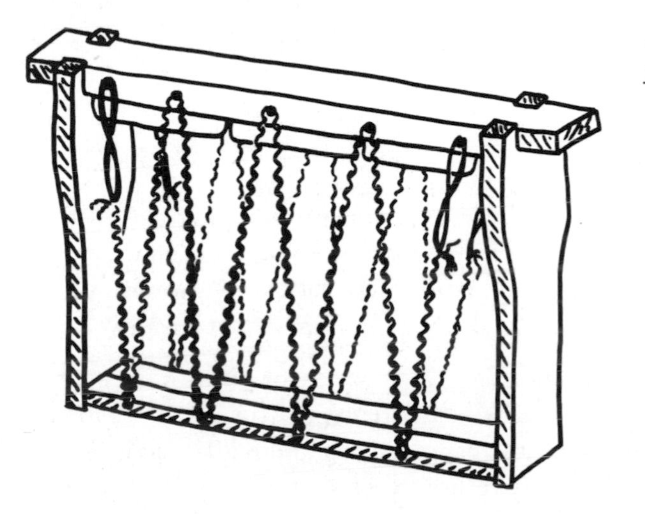

Dave Heilman's cutout frame.

You can also use large rubber bands by themselves to tie in comb. They are not good for the piecing together of small sections of comb, and in our experience, they are sloppy compared with the Heilman-style frame. Rubber bands have a big advantage in that you can keep them on hand (they take up no room), and put them into service with no preparation.

Making Splits from Existing Hives

Once you have a self-sustaining population of bees, getting more bees is relatively straightforward: you split the colony into two or more separate colonies. In some cases, you might even be able to make a split from a first-year hive. In general, the more resources you can give the split the better, but there are good reasons to make weaker splits as well. If you are performing a weak split, make sure you know why you are doing so and compensate where necessary.

The Necessary Resources

When doing any kind of split, make sure that each resulting colony has resources in the proper balance. This includes capped brood, capped and uncapped honey, nurse bees, foragers, beebread, and drawn comb. Uncapped brood requires nurse bees to care for it, honey requires guards at the entrance to prevent theft, and rearing a new queen requires an abundance of honey, pollen, and nurse bees as well as young open brood and, hopefully, eggs. If a split (or parent colony) ends up with open brood but no stores or nurse bees, that brood will likely die.

Smaller splits become nucs, and are important management tools. Nucs represent a reserve of bees and queens that can be used to requeen or boost an existing colony, or can be built up to increase the number of colonies one has.

Reproductive Splits

The most obvious reason to split hives is to make more hives. Called reproductive splits, these are generally performed on strong colonies when there are plenty of resources available, such as at the beginning of a flow. Wait until you have eight frames of brood to work with, and make the split with at least two or three frames.

The queen can either stay with the parent colony or go with the split. It's a good idea for the foragers to go with the half that doesn't get the queen. If you leave the queen with the parent colony, move the parent colony and put the split in the original location; the foragers will naturally return to the split.

One common and productive strategy is to split a strong colony just before a major nectar flow. Move the queen with the split, and let the parent colony in its original location make a new queen. Make sure to leave most of the capped brood with the parent colony, but not all of it; the split will need young bees as well. This makes a very strong honey production unit, as the energy required to raise brood has been diverted into honey production until a queen has been raised, she has mated, started to lay, and the eggs hatch, which will take about three weeks. In the meantime, the old queen has been building up the nuc, and by the time the flow is over it is strong and expanding.

You can save some time when making splits if you use queen cells that you find in your colonies, purchase queens for your splits, or use your own queens that you raise in mating nucs (see Chapter 11).

Splits as described here are great ways to make modest increases in small or large apiaries. Other more complicated methods for making splits are more efficient, but such techniques are beyond the scope of this book. Mostly, they reduce the amount of time before the queen in a newly made nuc begins to lay.

The Least You Need to Know

- Your goal should be to have locally adapted, untreated bees. If you can start with this, you are ahead of the game.
- Free bees are available from feral hive cutouts and swarms, but they are difficult for a beginner to obtain safely.
- Purchased bees are more predictable to work with.
- Bees in a functioning colony with comb stores and brood build up faster than bees in clusters.
- You can split strong colonies into two or more smaller colonies.

Chapter 5

Finding a Home for Your Hives

In This Chapter

- Providing a source of water
- Making sure forage is available
- Avoiding highly populated areas
- Keeping predators at bay
- Equalizing your apiary

Like many aspects of beekeeping, choosing a place for your bees to live requires some consideration before they arrive. How your bees interact with the weather, local plants, wildlife, and people will rest largely on where the bees are placed. The hardest part of making such decisions is that we can't always predict such things. Will the field flood in heavy rain? Will local teenagers disturb the bees? Will a neighbor complain? Will bees encounter pesticides foraging in the nearby agriculture?

The best one can do is to consider the possible upsides and downsides of any location, and pay attention to how things work out. Don't be hesitant to move your bees if you think there is a better choice of location, but don't fall into the trap where you are constantly moving hives chasing a "better crop." The grass is ALWAYS greener in another location, but you cannot secure a crop (or even keep the bees fed) by constantly moving, fishing for the magic spot.

Location, Location, Location

Keep these basic things in mind when choosing a site for your apiary.

Water

Bees need water, and the farther they need to fly to find it, the more energy the hive must expend to raise brood and cool the hive.

Once the bees have settled on a water source, they are likely to keep using it no matter what else becomes available. Keep this in mind when you first set up your hives. If your neighbor has a swimming pool, get the bees used to a different water source *before* they get a taste of the pool. Bees will generally not be aggressive when collecting water, but poolside activity can be unpredictable, and no one sitting at their pool wants your bees around.

Bee Aware

If you provide an artificial water supply for the bees you must keep it consistently filled. Otherwise the bees will find a new source and you will not be able to get them to go back to your artificial source.

Birdbaths, year-round streams and ponds, and livestock watering tanks are all good water sources, and a site with one of these options close by is preferable. If you need to provide a source of water for the bees, a barrel with floating twigs to give the bees a landing place (bees drown better than they swim), or a board with a trickle of water from a hose both work well.

Sunlight

Short of a crowing rooster or its urban equivalent, the car alarm, nothing gets us up in the morning earlier than rays of sunlight streaming through the bedroom window. The same is true for bees. Situating the hives so that sunlight falls on the entrance early in the morning is the best way to encourage your bees to get an early start to the day.

Nearby Food Sources

The availability of forage for the bees is of prime importance, but is difficult to determine ahead of time. Most possible locations will have abundant forage for some of the season. This is not a good determining factor, as it is the lows, not the highs, that dictate how many hives a particular location can accommodate (from 0 to 100 hives per location is the general range).

In times of abundance, there is going to be more forage available than your bees can collect. In times of dearth, however, bees will quickly consume their stores.

Bee Smart

If there is enough forage to sustain 20 colonies during the flow, but only enough for 1 colony in dearth, you have to assume that if 20 colonies are placed there, each can collect only 1/20 of what it needs to avoid depleting its stores in a dearth.

The best way to determine the proper number of hives is to start with a few, and monitor the progress as you make splits and grow the yard. When you see large hives losing weight rapidly in dearth, you know you should move some of them to a less populated location.

It's worth noting that insecticides and fungicides are widely used on agricultural crops and on lawns and gardens. Make sure your bees are in areas where they have access to untreated forage. Wild tracts of land with lots of weeds and wildflowers are best for bees. Other beekeepers in your area will know about local conditions (crops, wildflowers, and pesticide usage), and are an important resource for you.

One thing to keep in mind is that in areas where there are hills, mountains, and valleys, the same plant blooms in different elevations successively. Generally, it's warmer at the bottom, and cooler at the top, so the bloom starts down low and moves up. If bees can be placed in such locations, they can extend the flow of each source and can greatly increase the honey yield.

Wind

Dynamic air currents are important for bees. The hive has to be able to shed moisture to cure honey and cool the hive, mostly at times when the humidity tends to be high. Like humans, bees exhale water vapor, so bees working hard produce a lot of moisture. Avoid locations without air currents such as the bottom of a gully. Fresh, moving currents are essential.

Although circulating air is important, try to avoid situating your hive so that cold winter winds blow directly into the hive entrance. Take note of the prevailing winds when you select a location and orient the bees accordingly.

Bee Bonus

If you are in a more populated area, tall weeds and flowers do a great job of camouflaging beehives. Start a trend in your neighborhood and turn your cultivated lawn into a decorative weed patch. By mowing small areas or adding some trellising or yard art, you can make your weed patch look deliberate. Almost everyone wants to help the bees. See if you can start a weed revolution in your community. Your beehives, reconfigured or decorated, can even become yard art!

People

The flight paths of bees should not intersect where people often walk. Since flight paths can change through the season, it's best to do something to get the bees to fly up. Placing an obstacle such as a fence, hedge, or wall in front of the entrance will do the trick. Bees don't do much of their foraging near their own hives, so helping them get up into the air will not adversely affect them.

Bee Smart

Many municipalities have laws governing the number of colonies you can keep on a given plot of land, especially in urban areas, but there are very few places in the country where keeping bees is prohibited. Always check with your local town or city hall before placing bees.

Always have a backup location in mind for your hives. You never know when a property owner, wildfire, neighbor, or city council will require that hives be moved immediately.

Large Predators

Potential predators are a separate consideration when choosing a location for bees.

Bears

Bears are most often dealt with by using an electric fence baited with bacon strips or rag strips soaked in animal fat and tied to the fence wires. The bear will sniff, lick, or chew the electrified bait and be scared off before getting to the beehives. Once a bear gets a taste of honey, nothing will discourage it and you'll have to move the hives.

Bee Bonus

We know some beekeepers who deal with bears by hoisting their apiary into the trees with ropes and pulleys. This is a fascinating approach, and it works well, but it does add a whole level of complexity to things.

Skunks

Skunks are another common problem. Their tactic is to come at night, scratch at the entrance, and eat the bees as they come out to investigate. Scratched-up earth and some chewed-up bees at the entrance are good indicators of a skunk problem.

Anything that exposes the skunks' underbellies to stings when they scratch the entrance does a good job of preventing problems. Entrances that are high off the ground work well. If you use a traditional bottom board and have a skunk problem, place some chicken wire (2" or larger mesh) in front of the hive, from the ground diagonally up and over the front of the landing board. This will force the skunk to stand on its hind legs.

Mice

Mice often move into hives in the fall as the weather gets cold. They damage the comb, smell bad, and are not welcome. You can usually avoid a mouse invasion by placing a *mouse guard* over a bottom entrance in the early fall, before the mice move in and do their damage.

def•i•ni•tion

A **mouse guard** is a barrier that, when placed over a hive entrance, allows bees to pass through but is too small for a mouse.

Sometimes bees will kill a mouse in the hive but are unable to physically remove it. When this happens, the bees will coat the mouse in a layer of propolis. Composed of plant resins collected by the bees, propolis is essentially turpentine and is extremely antimicrobial. A mouse sealed in propolis doesn't smell or ooze all over the hive.

Birds and Dragonflies

It's perfectly normal for birds to eat your bees on occasion, but if you have a large population of bee-eating birds it can start to impact your hive populations, or worse, pick off the big, juicy, slow-flying queen. Similarly, yards located near water may have a large number of dragonflies, which also like munching on bees. Such locations can be great honey producers, but terrible places to rear queens.

Equalization

Equalization is the process of transferring resources from stronger hives to weaker hives. This is done to keep colonies in a specific location at similar strengths to prevent robbing and to secure a large honey crop.

Once you start to acquire more hives and keep them in more locations, you'll start to think by the yard rather than by the hive. Each yard, each location, has its own set of circumstances. An unproductive location can be less than a mile from a productive one. In such a case, it's not "how many" that counts, it's "where." Only by keeping hives in more than one location can you learn what works best in your area. Weather, climate, predators, forage, pollution, human activity, and geography all play into the strength of the yard.

Within each yard, you should have two goals with regard to hive strength. First, you want to find out which are the strongest hives so that you can use them in your breeding program. You also want to keep all the hives in the yard built up enough to provide a honey crop. These goals are perhaps a bit conflicting, but not impossibly so. There are two primary ways to equalize hives within a yard: you can transfer frames or move hives.

Transferring frames is a very direct method. From a hive that has more than eight frames of brood, transfer a couple of frames of capped brood (the closer to emerging the better) to a hive that could use a boost. Similarly, if a hive is starving, feed in frames of capped honey or ripening nectar.

Bee Smart

Take care not to waste honey and/or brood on hives that are too far gone to help. If a hive is really hurting, don't spend too many resources on fixing the problem. Use its resources to boost other hives instead of weakening them to prop up a loser. If it is not too late in the season, starting a nuc from a stronger hive is usually more productive than propping up a weaker one.

Forager bees will return to the location they are oriented to. If you switch locations of a weak hive and a strong hive, the large foraging force from the strong hive will boost the weak hive, and the strong hive has resources to assign more bees to foraging. Do the switch when the foragers are out flying. Moving hives is less invasive than swapping frames and really works well.

The Least You Need to Know

- Locations can make or break your colonies.
- Bees need abundant, untreated pollen and nectar forage.
- Place hives near a good water source and make sure they have plenty of ventilation.
- Think by the yard, not the hive.

Chapter 6

Bees Are What They Eat

In This Chapter

- The bee diet: pollen and nectar
- Letting bees keep their honey
- Unnatural feeding practices
- Breaking the brood cycle

It's easy to overlook the bees' nutritional needs because, in general, they feed themselves and choose their own diet. Bees do best when they have a variety of pollens and nectars to forage upon. Unfortunately, conventional beekeeping practices often involve altering the bees' diet. For instance, moving bees for pollination contracts (especially in monocrop plantings) restricts the bees' diet to perhaps one source. In addition, beekeepers often feed their bees low-cost sugars (sucrose, high-fructose corn syrup/HFCS) for overwintering in order to extract and sell more honey (which is worth considerably more on the market than the cost of the sugar feed).

With these practices in mind, let's look at how bees feed when kept in a more natural system.

The Honeybee Diet

Proper nutrition is critical to honeybee health. Unlike many other insects, honeybees are dietary generalists and require a diverse range of pollen and nectars.

Protein and Carbs

Pollen is the protein source for bees. It is necessary to build strong muscular and glandular systems. Nectar/honey is the carbohydrate source that keeps the bees running once they are built. Adult honeybees don't build or repair tissues, they simply refuel with carbohydrates to "run the engine," until they literally wear out.

As the bees gather pollen, they moisten it with saliva and nectar carried in their honey stomachs so that the pollen will stick together in transport. These secretions and the pollens themselves contain microbes that immediately (even before the bee has returned to the hive) start the fermentation process, which transforms pollen to beebread.

Bee Bonus

The bees pack pollen in what are essentially "baskets" on their hind legs to carry it back to the hive. Each leg has a single long, stiff hair perpendicular to the leg, which gives the packed pollen support for the ride home. When the pollen flow is heavy, bees approaching the entrance look like they are wearing brightly colored saddlebags.

Fermentation

When the bees return to the hive, they pack the pollen into cells, where the fermentation process continues. Over several weeks, intense microbial activity occurs and vitamins, enzymes, proteins, and other essential nutrients are created and/or released. It is important that this microbial succession occur uninterrupted, as each step of the fermentation process sets the stage for the next, and produces its own unique set of byproducts. Even after the fermentation is complete and the beebread is preserved, the process isn't over. Enzymes continue to be active for some time after the pollen is fermented. Beebread has been found to have twice the amount of water-soluble proteins as raw pollen.

Young adult bees consume protein-rich beebread to complete their own development, produce bee milk, and feed larvae directly. The hypopharyngeal and mandibular glands in the young bees' heads turn the beebread into protein-rich royal jelly for the queen larvae and brood food for the worker and drone larvae.

If there is even a temporary shortage of pollen, the brood may be undernourished, affecting their muscular and glandular development. Since these resulting bees will be responsible for nursing and foraging to provide for following generations of brood, the repercussions of a lack of pollen can persist for months.

Because the brood turnover time is so short (about three weeks per cycle) and the queen can lay so prolifically (up to 3,000 eggs per day), nutritional deficits can severely impact the hive.

Honey

Honey, concentrated by the bees from nectar, contains sugars as well as a complex array of acids, enzymes, proteins, amino acids, vitamins, minerals, and mineral salts. The bees need these diverse nutrients.

Forager bees collect nectar and store it in their honey stomachs. When the foragers return to the hive, house bees take the nectar from them and begin the process of transforming it into honey.

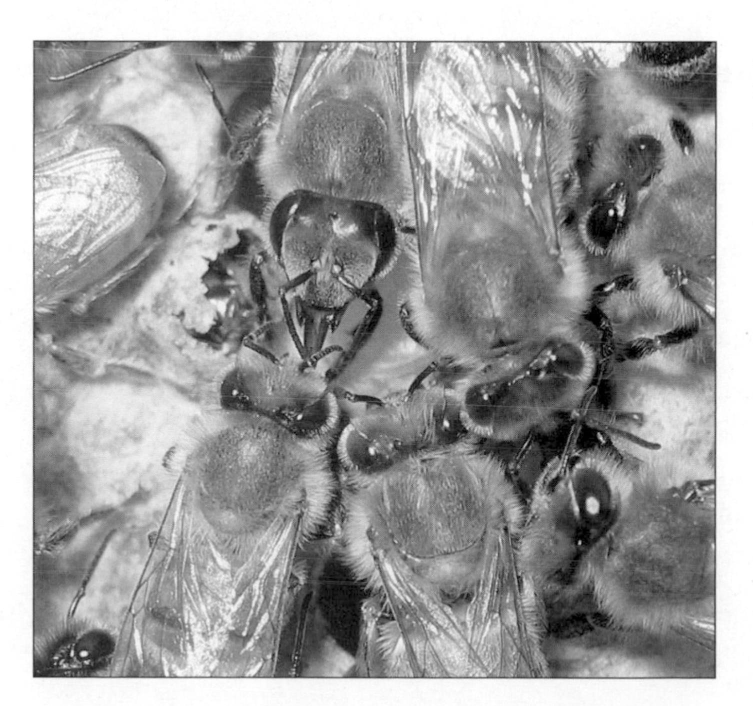

A forager uses her tongue to transfer nectar to a house bee.

Some of the work has already begun, as microbes in the honey stomach have secreted enzymes that perform some of the more mysterious parts of the transformation. We don't know all the details of the process, but it's recently become clear that honey is actually a fermented food.

It takes quite a bit of energy to concentrate the nectar, which starts out between 30 and 70 percent water. The house bees fan over the nectar with their wings, generate body heat, and manipulate the nectar droplet by droplet with their mouth parts in order to expose as much surface area of the nectar as possible to the air.

When the moisture content is reduced to about 18 percent, the honey is more or less immune to spoiling (there isn't enough moisture for fermenting microbes to be active), but it is still vulnerable as it will attract water from the air until it becomes dilute enough to sour. At this point, the house bees build wax caps over each cell of ripe honey, preserving the honey inside indefinitely.

Honey is the fuel that bees run on year round. At the height of the season, when there might be in excess of 60,000 bees in the colony, you need to remember that the honey being stored is only the excess that the bees are not using to eat, fly, concentrate honey, and raise brood. Honey allows the bees to overwinter, as well as to maintain a large population through seasonal and weather-related dearths.

Manipulating the Numbers Through Feeding

Much of the beekeeping literature and conventional management practices are geared toward micromanaging the number of bees in a hive. This is often achieved by feeding the bees high-fructose corn syrup (HFCS), cane sugar solution, and pollen supplement when there is a dearth or the bee population begins to wane due to lack of available forage.

Common feed management practices include:

- Feeding bees sugar, high-fructose corn syrup and pollen supplement before a flow to increase the foraging population.
- Feeding sugar, high-fructose corn syrup and pollen supplement every time the population declines, instead of letting the bees' population ebb and flow with the season.

Although these techniques work to a greater or lesser degree, they all interfere with the natural functioning of the hive. We feel that such interference, no matter how well intentioned, is generally harmful for the bees' long-term survival and should be avoided whenever possible. Most importantly, it prevents the bees from doing what they do best, which is to adapt their behavior in response to conditions in the world around them.

Accept No Substitutes

Let us be very clear about feeding bees high-fructose corn syrup and other sugar solutions: these are *not* acceptable substitutes for honey as bee food. Their nutritional values are not equivalent. They also do not have the same pH as honey and so alter the microbial culture of the hive. Many bee pathogens grow more readily at the pH of sugar syrup than at the pH of honey.

Note that pollen supplements are just that: supplements and not substitutes. The complete nutrition that real pollen provides cannot be substituted by any man-made recipe. If the bees are without real pollen for only a few weeks they will begin to show nutritional deficiencies, with or without the supplements. Although pollen substitutes may at first seem to increase worker populations, the resulting bees may be of a lower quality with shorter life spans.

Honey and pollen are the bees' food. Bees do not make cane sugar, HFCS, or pollen supplement for their consumption; they make honey and ferment pollen, and they should be allowed to have real honey and pollen whenever possible. There are no equivalent substitutes.

Even if you think that sugar syrups and HFCS are appropriate substitutes for honey in your own food, do your bees a favor and let them have what they are intended to eat.

Bee Smart

Beekeepers who feed sugar syrups and/or HFCS to their bees sometimes claim that these products do not migrate into the honey intended for human consumption. To prove otherwise, we suggest a small experiment: add blue food coloring to all your artificial feed and see what happens. Bees move materials from one part of the hive to another all the time, and colored syrup is no exception. Feeds often get stored (and harvested) as honey.

The natural functioning of the hive has kept bees around for 100 million years without our help. Interfering with such important processes of reproduction and population can provide short-term gains, but only at the cost of long-term and continued adaptation to a constantly changing environment. Our interferences suppress the bees' ability to care for themselves without our help. Most of all, these interferences prevent the bees from having an accurate perception of what is actually happening around them and reacting appropriately.

Limiting Factors

Food isn't the only resource needed to raise bees. Obviously, a queen and comb are required so that eggs can be laid, but the limiting factor is more likely to be the number of nurse bees.

A good rule of thumb is that it takes one cell of honey, one cell of pollen, and one cell of water to make a bee. Just maintaining the population and stores requires an almost unimaginable amount of all these resources.

A colony can only raise the amount of brood it can afford to feed and keep warm. The beekeeper can, of course, feed the bees. Keeping the bees warm, however, involves some very simple rules of math and geometry that humans have not been able to overcome.

Picture a brood in the comb. Each cell contains one developing bee. Bees keep brood warm by covering them. Now look at the relative sizes. Each adult bee can cover two brood cells with her body. This is the limiting factor.

One adult bee can cover two cells.

Based on this simple math—one adult bee to keep two brood cells warm—we can quickly determine that the bees can, at most, raise about twice the number of brood as they have bees to care for them. Since each worker takes about three weeks to emerge, in ideal conditions you can count on the bees doubling the number of brood cells every three weeks. In three weeks the first cycle of brood will all be available as nurse bees. No matter how much you feed the bees, you cannot change this math. If you have 3 frames of brood today and conditions are ideal, you will have around 6 frames of brood in three weeks, 12 frames of brood in six weeks, and so on. Of course, this assumes there is fully built comb to occupy, favorable weather, and sufficient forage. If bees must first draw comb, it will take longer for the population to double in size.

Your role as beekeeper is to make sure there is room for expansion of brood or stores so that the bees can regulate their own population throughout the year. This is essential for bees so that they can build up quickly in anticipation of a flow.

Stores vs. a Flow

A honeyflow stimulates bees, encouraging them to rear brood. They feed on the honey prodigiously and expend great amounts of energy taking care of the brood.

Bees respond differently to the presence of stored honey, especially capped stores, than they do to a flow. They treat stores as we might treat canned goods in the pantry: as a long-term supply of food to use in times when the garden isn't producing or when the snow is too deep to get to the store. This is not food for a feast or party, and it is not seen as enough food to stimulate brood rearing. It is stored energy, future sustenance. For this reason, you should leave plenty of honey with the bees at all times.

Feeding bees high-fructose corn syrup, cane sugar, or even honey has the same stimulative effect that a honeyflow does. It communicates "there is a flow on" to the bees, which encourages brood rearing.

The bees should be stimulated when nature stimulates them, not when we decide to feed them.

Managing for Long-Term Success

Bees that can keep enough honey to continue reproducing are bees that require minimal intervention by the beekeeper. This is the baseline that must be achieved *before* you can harvest any significant honey crop. Building the colony, the yard, and the apiary are the foundation of everything to follow.

Beekeepers must resist the temptations of short-term gains—a bigger honey crop and more colonies this year than last—and instead work for long-term sustainability. If we don't, we commit ourselves to ever-declining health of our bees and an unbreakable reliance on artificial treatments.

Breaking the Brood Cycle

Constant feeding has another negative consequence for bees. Remember that a single brood cycle is three weeks long (the time it takes to double the size of the active broodnest under ideal conditions). It is also the time it takes for the youngest eggs to hatch into brood, and to emerge as adult bees.

Feeding bees encourages bees to continue rearing brood even though they would otherwise cease doing so. This can have disastrous consequences for the health of a hive.

Bee Smart

We consider a new colony from a package or nuc to be an exception to the general prohibition on feeding. A new colony with few or no food stores must sometimes be fed to give it a chance to survive. This should be a temporary solution only until the bees build up enough stores to survive on their own. (See Chapter 7 for details.)

Disease and parasites often reproduce and sustain their population in the brood. When bees stop laying and rearing brood, they interrupt the life cycle of diseases and parasites. This is one way that bees living on a natural system (and in nature without the "help" of a beekeeper) sustain themselves. Breaking the brood cycle can eliminate diseases and parasites. Bees also break the brood cycle when they swarm.

This is all to point out that the practice of feeding can interfere with some of the bees' natural defenses, and should be considered as an emergency measure, and not a regular management practice.

The Least You Need to Know

- Bees are meant to eat honey and pollen.
- Honey is produced through the concentration of the nectar, and through fermentation.
- Feeding bees high-fructose corn syrup and other sugar solutions alters bees' natural breeding cycles and changes the microbial balance within the hive.
- New colonies often require feeding, but it should be done only if absolutely necessary.

Chapter 7

Establishing Your Hives

In This Chapter

- Installing packages and nucs
- Deciding whether—and when—to feed
- How to feed your bees
- Adding boxes to your hive
- What to look for in your hive

Once you have your bees, whether from a nuc, package, swarm, or cutout, you need to put them somewhere. This chapter walks you through the steps of introducing your bees to their new home.

After the bees are installed, you'll need to monitor them closely to make sure they are adequately preparing for the winter. Although we typically discourage beekeepers from feeding their bees, newly established colonies are the exception.

Hiving Package Bees

Your package should be stored out of the sun. Too cool is better than too hot, as it's easier for the bees to cluster and keep warm than it is to fan and keep cool when confined in the package.

Until you open the package, you have no way of knowing the status of the feed in the can. If the holes are too small, the bees may have had trouble getting enough syrup out. If the package was mishandled or caged up for a long time, the feed can may be empty. To ensure your bees have enough food, spray them with a thin sugar or honey syrup through the cage from time to time until you transfer them to a hive. Don't soak them, but mist them as frequently as they will take the feed. Don't apply the feed with a brush, as doing so can damage bees' feet and tongues.

Before hiving the package, you should have the hive and frames in place and ready to go. Late afternoon is ideal for installing packages, as the impending nightfall will encourage the bees to stay in the hive, and it will smell like home by morning.

Bee Aware

Unlike most work you will do with the bees, when hiving a package, a smoker is unnecessary. If you are hiving a number of packages, you may choose to keep a lit smoker nearby in case of stings. Smoking the sting site will help mask the smell, preventing other bees from getting riled up.

Step-by-Step Instructions

The procedure to hive a package reads like a fraternity hazing prank. We swear this is what you are supposed to do:

1. First, examine the package. You should see a box full of bees that is screened on four sides. A few adhering bees might be clinging to the outside of the screen, but bees should not be leaving the box.

   In the middle of the box is a can full of sugar syrup. Hopefully holes were punched on the bottom of the can (yes, the bottom … don't worry, gravity and vacuum will keep the syrup from pouring out as long as the package remains right side up).

   Next to the can of syrup is the queen cage, which is hanging on a metal tab held in place by the can. Bees will likely be clustered around the cage and the can.

   The can is held inside the box with a wooden support structure, and a small sheet of wood or fiberboard is stapled over the top to retain the can and to keep the bees inside.

be located for a day or two and open an entrance. (Not all nuc boxes are designed to be placed outside, so use common sense here.) This way, when you install the nuc, the foragers will already be trained to return to their hive location. This will prevent bees from getting disoriented during installation.

Bee Smart

If you had previously placed the nuc where you plan to place the hive, move the nuc off to the side several feet and position the new hive body in its place. Notice that the nuc is easy to handle, as the foragers are flying around the new hive, not the nuc. If you haven't oriented the nuc to the location, open the nuc box as close to the hive as possible. This way, as bees fly out they will find the hive and the queen quickly.

Since the bees in a nuc are already on frames, hiving them is easy—well, as easy as opening the nuc, pulling out frames, and placing them in your new equipment. Of course, if this is your first time handling bees, it isn't so simple. Here's what you do:

1. Start by putting the hive in place, and making room for the nuc frames. Leave an extra frame out of the hive for now, as it's easier to manipulate frames when there is extra room. Now is also a good time to get your smoker lit, as some gentle smoking as you get started will help keep the bees calm.

2. Use your hive tool to pry up an outside frame from the nuc; pick the one with the least visible activity. Once the first frame is out, the rest will be easier to handle. Make sure to retain the order and orientation of the frames when you transfer them from the nuc to the new hive.

3. Place the nuc frames in the center of the larger box. As you place frames in the box, make sure that the side bars slide against and touch the side bars of the adjacent frame. If bees can't get in between the side bars, they can't get squashed.

4. Before closing the hive up, look to see if any bees will be squashed when you put the top on. Use your smoker to gently drive the bees off the box edges and down between the frames as much as possible. Too much smoke will agitate the bees, so keep it to a few light puffs.

As you transfer the nuc, you should see at least one frame with honey. If there is very little honey (less than a quarter of a frame) and there is no significant forage available (consult other beekeepers in your area), you should consider feeding the bees.

Bee Aware

If circumstances permit, always begin and end any intrusion into the hive by pulling a frame where there are few bees. There is always a chance that a queen will be rolled (crushed by the bees) and injured or killed when pulling frames, so pay close attention when doing this. Always look down at the box before pulling the first frame. If one area looks less populated than the others, start with a frame in that area. Always insert the final frame into the least active side of the hive.

Some nucs are shipped with queens in cages, and you should follow the instructions provided by the supplier for releasing her.

To Feed, or Not to Feed?

Whether to feed or not really depends on circumstances. You should always feed package bees, as they have a lot of work to do before they can even begin to be self-sufficient. Even though a package hived on pre-formed plastic comb such as HSC will not need to draw comb, it takes time for the bees to accept the comb, and feeding helps this process along.

Since a nuc is already a functioning hive, as long as the bees have a small amount of stores and available forage, they can begin taking care of themselves right away.

Your local bee club or an experienced beekeeper in your area is your best guide as to when the flows and dearths tend to be, and will be helpful in figuring out if and for how long you need to feed your newly hived bees.

Conflicting Needs

When it comes to feeding a new colony, you'll have to reconcile some conflicting needs.

The moment you install bees, your goal is to help them prepare for winter by drawing comb, storing food, and creating a large population of healthy, well-nourished bees. This requires energy. If they don't have enough of their own stored food and there is a dearth, the bees may not have enough energy to do what they need to do to survive *unless you feed them*. When this is the case, it makes perfect sense to feed your newly installed bees; otherwise, they will die. However, you need to be careful not to overfeed, as doing so will discourage foraging.

Keep in mind that in most locations, bees do not constantly build up and raise brood throughout the season. Instead, the bees react to and even predict fluctuations in available forage, and regulate their populations accordingly. Breaks in the brood cycle are natural, and help the bees deal with diseases and pests, especially those that rely and/or feed upon brood. You should not automatically feed bees if there is a break in the brood cycle.

Bee Aware

If you let the bees build comb without foundation (using foundationless frames or a top bar hive), be extra careful about feeding. Feeding simulates a flow, and the bees are apt to draw large honey storage cells in excess, and may end up with limited worker brood comb.

Sometimes bees will go so far as to eat eggs and young brood. The prevailing wisdom is that in such circumstances, the bees must be fed; however, we do not subscribe to this theory if the colony remains strong. Bees are beautifully designed to adapt to their local conditions and should generally be allowed to do so.

What to Feed Your Bees

Despite our reluctance to encourage feeding, it is sometimes necessary. When it is necessary, you need to decide *what* to feed your bees.

Our first choice is clean honey. This means honey from a known beekeeper that does not have a disease problem and, more importantly, does not use antibiotics in their hive. Antibiotics disrupt the microbial culture of the hive. We cannot stress enough that cheap honey from the supermarket will not cut it here, nor will honey from a beekeeper who uses antibiotics.

Assuming clean honey isn't readily available—it isn't in the vast majority of cases—cane sugar made into a syrup is your best bet. Use one part sugar to one part water by weight (1:1).

It takes a great deal of energy to evaporate enough water out of nectar so that it won't ferment. Nectar coming into the hive stimulates brood rearing and comb building. Feeding 1:1 syrup has a similar effect. This is the type of solution you should feed a package or a nuc to give them a boost, as you want to provide food and encourage them to build comb and raise brood.

Feeding 2:1 (two parts sugar to one part water by weight) doesn't have quite the stimulative effect. A 2:1 syrup solution requires less energy to ripen and store, and is used for fall feeding to prepare colonies without sufficient stores for winter. Feeding 1:1 in this case simply adds to the amount of work the bees must do before the feed can be stored.

There are a number of ways to feed the bees, and the best choice depends on circumstance.

Feeding Package Bees on Honey Super Cell (HSC)

For feeding package bees that you've installed on HSC, there really is only one good way. Directly above the box of HSC you should place either a standard inner cover or a board with a 1½-inch hole in the center. There should be bee space (if using an inner cover) or a little less (if using a board with a hole in it) between the board and the tops of the frames.

You'll need a plastic pail with a tight-fitting lid or a brand-new clean paint can of up to two to three gallons.

1. In the center of the lid, drill about a dozen holes ⅛ inch in diameter.
2. Fill the pail with syrup, making sure the pail is mostly full, otherwise the vacuum will not form properly.
3. Secure and seal the lid on the pail and invert the entire pail over the board centered over the hole.

   This should give the bees access to the syrup, without allowing them to escape around the rim. Although a little syrup may drip at first, the vacuum formed will hold the bulk of the syrup inside the pail until the bees remove it with their tongues.

Bee Aware

If you are checking on bees being fed with a pail feeder, take care not to turn the feeder over. If there is not enough feed left in the pail, it might not be able to re-establish a vacuum without adding more feed.

4. Place an empty hive box around the pail and secure some kind of cover over it to prevent raccoons or other animals from tipping it over and to protect it from the sun.

This kind of feeder will keep the bees clustered below it, and in the case of HSC, between the frames. It also works well with other hive configurations, especially when starting from a package, which will require more feed than a nuc.

Feeding a Nuc or Full-Size Hive

If you want to feed a nuc or a full-size hive to help them along, you have a couple of good options. You can do as described above, or you can use a baggie feeder. Baggie feeders work in most circumstances; the primary exceptions are if you need to feed frequently, or if you are trying to get a queen to lay in HSC (the bees will build comb between the top cover and the top bars). To install a baggie feeder, follow these instructions:

1. Construct (or buy) a shim the same dimensions as the hive body and about an inch tall. Remove the top cover(s) and place the shim directly on the topmost box. The shim is there to make room between the frames and the cover for the baggie feeder.
2. Fill a 1 gallon sealable plastic bag to about three-quarters capacity and seal it. You can do this ahead of time.
3. Gently smoke the bees off of the top bars. Don't overdo the smoke … a few puffs over the tops of their heads should drive them down into the comb.
4. Lay the bag sideways directly on the top bars and then cut a large "X" into the bag with a sharp razor blade. A dull blade will make this difficult, so use a brand-new one. Amazingly, the syrup mostly stays in the bag, and the bees slurp it up from the slits.

Feeding Pollen

Occasionally a hive may need to be fed pollen. This can happen during particularly bad seasons when forage is scarce or when a new colony hasn't been able to build up its numbers to store excess pollen before winter. In both cases, the hive may need some pollen during the winter to jump-start brood rearing before the first spring flow.

Take note of the pollen stores going into winter. If pollen in the hive is scarce, you can swap in a frame of pollen from a hive that has some to spare or that you plan to feed. Remember that it takes a cell of pollen, a cell of honey, and a cell of water to build a bee.

First, you need to source clean pollen and honey (preferably crystallized). All beekeeping journals have ads for pollen suppliers, and there are sites on the Internet for pollen as well. You may also find contacts within your local beekeeping community. Contact suppliers and ask the following questions:

- **Where does the pollen come from?** If it's foreign it will probably be irradiated, which will have killed the microbes necessary for proper fermentation.
- **Was the pollen gathered from bees foraging on agricultural crops or wild land?** Do your best to obtain the cleanest pollen and honey you can.

Once you have the pollen and honey, mix them together into a thick paste and form the paste into patties, as if you were making a burger. Sandwich the patties individually between pieces of wax paper. Make a few slashes in the bottom of the paper and using the shim, lay a patty on the top bars of each hive you are going to feed, and cover the cluster.

If no honey is available, you can mix the pollen with dry sugar, and add just enough water to make a paste.

Building Up the New Colony

In the case of packages, you have to face the fact that they will not grow right away. Ideally, packages are made up of mostly young bees, but some older bees are present, and some bees will die in the package.

When you install package bees, *at least* three weeks will pass before any new bees emerge in the colony (even longer if you're using HSC). In the meantime, bees will be dying of old age and from predators in the field.

Initial Buildup

If you can get the bees to build out comb (or start storing syrup and continue laying in HSC) while the first brood is developing, you are doing well. Comb is the essential resource, and you must monitor it carefully. Without sufficient comb, the bees will not be able to store food for the winter or build up their own population to sustainable levels.

Bee Smart

In general you should avoid moving frames around. Let the bees structure their broodnest as they please.

Most importantly, the bees put the brood all together so that they can be cared for. Do not separate the broodframes from the broodnest. If you want to expand the broodnest when it is buttressed by frames full of honey, add a foundationless frame or a frame with foundation only, putting it just outside the existing brood. Only do so when the population is large enough that when you remove a frame, bees on either side start to fill in the gap. Without a high enough population density, adding frames stresses the bees unduly and won't help them draw out new comb.

Adding a Box

Once the bees have covered 8 frames in a 10-frame box or 6 frames in an 8-frame box, it is time to add a second box. No matter if your frames have foundation or are foundationless, you should use *bait comb.*

Follow these steps:

1. Gently smoke the entrance and open the hive. Place the new box off to the side with the center three frames removed. These frames can be either foundationless or wired with small cell foundation.

2. Remove an outside frame from the colony to make room for manipulations. Transfer the three center brood frames one at a time to the center of the new box. Be sure to keep an eye out for the queen, and avoid losing her in the shuffle. If you see her on a frame, carefully replace the frame, and choose another. She is safer in the lower box while you are moving things around.

def•i•ni•tion

Bait comb is used to encourage bees to expand into a new box. Comb that is currently being used by the bees is swapped with empty frames (or combs) from the box being added. The smell of brood and stores will attract the bees into the new box.

3. Close the gap you've made in the center of the old broodnest by pushing the frames on the outside of the broodnest toward the center, and then add the empty frames from the new box around the sides of the broodnest. Replace the original frame you removed.

4. Gently smoke the bees down between the frames. Add the second box (make sure the frames are butted up against each other), and cover the box.

Whenever you add a box, you should follow this procedure, called pyramiding, as it helps the bees cluster at the top and draw comb down and out.

The Least You Need to Know

- Don't drop the queen cage into the package.
- Make sure the screen on the queen cage is perpendicular to the comb.
- Don't overfeed.
- Bait the bees by "pyramiding up" the frames when adding boxes.
- Keep the frames in the order the bees work them as much as possible.

Chapter 8

Inspecting Your Hives

In This Chapter

- Inspecting the inside of the hive
- Inspecting the outside of the hive
- Identifying diseases and parasites
- Common comb problems

In quantum mechanics, there is a concept known as the Heisenberg Uncertainty Principle. One aspect of the principle that seems to resonate into all areas of science (and human experience, for that matter) is that the more closely something is observed, the more the act of the observation itself affects the subject of the observation. To see inside an apple, you have to cut it open, thereby altering it. A child acts differently when she knows she is being watched by adults. With bees, this "observer effect" is extremely pronounced. In order to "see" what the bees are up to, you have to take their home apart, which disrupts what they are doing … but what you really want to know is what the bees were doing before you opened the box!

Although bees do better when largely left alone, there are reasons (including our own curiosity) to inspect a hive thoroughly.

Performing a hive inspection is a skill, one that will become second nature through practice. With experience you will learn what a good hive looks like and realize when you don't need to go through every frame. But as a beginner, have at it—it's the only way to learn.

Inspecting the Inside of the Hive

Always conduct your inspection from the back of the hive, opposite the entrance, so you're out of the path of entering and exiting bees. Follow these steps:

1. Direct a couple of small puffs of smoke into all entrances, being careful not to overdo this.
2. Set up/plan for something to place the boxes on as you unstack them: an empty hive body, a top cover, a hive stand, or even the top hive body on end (on the end, not the side, so that the comb is perpendicular to the ground).

Here, the top box is laid on its end, and the other boxes are stacked on top. Note the orientation of the frames in the (now) bottom box. In this position, the comb will not collapse.

3. If you're using a telescoping cover, pull up one edge. If there is an inner cover with a hole, blow two puffs of smoke under the telescoping cover and close it back up for about 15 seconds before removing it completely. If the inner cover has no hole, remove the telescoping cover and proceed to the next step.

Bee Aware

When stacking boxes temporarily for inspection, always offset the boxes diagonally from one another, as the less contact area there is between boxes, the less chance there is of a bee getting squashed. You will squash some bees no matter what you do, but your goal should be to minimize the damage.

4. Using your hive tool, pry off the inner cover, which bees will tend to "glue" to the box with propolis. Pry gently to prevent the propolis from snapping all at once. As you get one edge free, direct a small puff of smoke inside. If there are a lot of bees present, close the inner cover back up and give the smoke 15 seconds or so to do its work, and then remove the inner cover.
5. Your hive tool will help you separate even the heaviest of boxes with the proper technique. You must lift the box from the back (the end closest to you) while prying at the back corners with a hive tool. Sometimes the top bars of the frames in the box below will be attached to the bottom bars of the frames above. When you pull the top box up, you'll pull up the frames from the bottom box as well. If this happens, use the hive tool to pry the bottom frames back down. When the box is free (but still resting on its front edge), lift it straight off and place it either on the stand you have prepared or on its end. Remember to keep track of the order and orientation of boxes and frames so you can replace them properly.
6. Don't mess with frames or bees yet. Pry off one box at a time, and stack them temporarily until there is only one box left on the bottom board. You will start your inspection at the bottom.
7. Inspect the bottom board by prying and lifting up the back of the bottom box. Put the bottom box aside if the bottom board requires closer inspection or cleaning.
8. Put the bottom box back in place, and look down on the top bars. Are there bees on all of the bars? Can you see them between the frames? Can you see the top rows of cells? Nectar? Brood? Capped honey? Uncapped honey? Pollen?
9. Find the frame with the least visible activity (usually one of the side frames). Use your hive tool to separate the side bars from the adjacent frames, and pry the frame up so that you can grab the top bar. Repeat on the other side of the frame so that you can gently lift the frame up and out of the hive.

10. Once you have created a space by removing the frame, you can manipulate things as needed. Whenever possible, keep the side bars butting against one another. Every time you have to push frames together, you run the risk of squashing bees. Keep the frames in contact with one another as much as possible, and just leave yourself space to work with the frame in question. Pull one frame at a time to inspect. As you return each frame to the box, push it away from the next frame to be inspected with the hive tool.

11. When you have inspected all the frames in the box, use the hive tool to push them all over so that the first frame can be replaced in its original position.

    Make sure that all the frames are pushed together and centered so that all of the side bars are tight against one another. This ensures that proper bee space between the combs will be maintained.

12. If there are a lot of bees on the top bars, use a little smoke to drive them down, replace the next box, and continue the inspection, working your way up until you finally replace the top cover.

What to Look For

As the season progresses, your colony of bees should grow. You'll need to inspect the hive regularly both to stave off problems and to advance your own learning about bees. As a beginner, your need to learn and observe outweighs the bees' need not to be disturbed. You can learn so much from observing your hives the first few years that we can't in good conscience suggest that you stay out of them.

Inspecting the Colony

Plan to inspect your colony every two weeks or when you suspect opposite there is a problem. Watch for the following things:

Eggs. Admittedly, eggs are small and hard to see until you get used to seeing them (and even then, in some light it is impossible). Eggs resemble tiny grains of rice, with one end adhering to the bottom of the cell. If you see eggs, you know there was a laying queen at most three days ago.

Open brood. When the brood is only a day or two old, it is almost as hard to see as an egg. Young brood floats in a pool of bee milk, and sometimes you see the sunlight reflecting on the surface of the jelly. As the larvae grow over a six-day period before being capped, they look like tiny, curled white shrimp expanding to fill the cells.

Bee Smart

Our friend and fellow beekeeper Michael Palmer points out what is rarely mentioned when inspecting a hive for brood: you aren't just looking for capped brood, open brood, and eggs, but a continuum of brood of different ages ranging from eggs to capped and even emerging brood all on the same frame, and in a visible progression. All things being equal, the queen should spiral through the broodnest laying eggs in a regular and fairly solid pattern. Things are not always equal, and just because you don't see this, it doesn't mean there is a problem, but there is a reason. If you are going to open the hive, you should be curious enough to try to find the reason.

Capped brood. This is one of the best ways to judge the strength of the hive. Capped brood represent bees that are past the stage of being groomed and fed; short of being damaged by sudden temperature changes or by a sudden drop in hive population, their presence means that there will soon be fresh, young bees that can raise more brood, build comb, and forage. Many of the bees you see flying in and out of the hive will be dead in a week, but plentiful capped brood is a sign of strength at least three to four weeks into the future.

The queen. It is not advisable to look for the queen every time you open the hive. Finding eggs is sufficient evidence that there is a laying queen, and spending the time to look through all the combs for her puts unneeded stress on the bees. It's nice to see her from time to time, but if you want practice spotting a queen, an observation hive is the best way.

Queen cells. You might see queen cells at different times of the year in the hive. There are really two reasons bees make queen cells: to supersede (replace) the queen and in preparation for swarming. Be careful to differentiate between unused queen cups with no egg or larvae inside and a queen cell that is in use. See the following section for details on how to handle the presence of queen cells in your hive. A queen cup requires no action.

Stores. Both honey and pollen stores should be present. Noting their fluctuations can tell you a lot about how the colony is doing. Abundant stores indicate a strong colony with resources to deal with challenges and dearths.

Troubleshooting

Throughout the first season, you should see comb drawn out and filled with bees, brood, pollen, and honey. Sometimes, however, things go awry. Here are some common problems and how to address them:

If you find no evidence of eggs or brood: A number of things might be going on. The most concerning is that for some reason there is no longer a queen in the hive. As insurance, you can add a frame from another hive with young open brood and eggs. If the hive is queenless, the bees will start to build queen cells and solve the problem themselves.

It's also possible that there is a dearth (especially if there are scant stores), in which case you should keep an eye on things and be prepared to feed if the situation gets dire. The broodnest could also be filled with honey and pollen with no room for the queen to lay. In such a case, the broodnest should be opened up with empty comb, foundation, foundationless frames (possibly in conjunction with adding a box), or by extracting the honey from the combs and replacing the extracted frames.

If you see a spotty ("shotgun") brood pattern: This, too, can mean a number of things. A new queen just starting to lay may take a few weeks to rev up, so don't judge her too quickly. If there isn't room for honey storage outside the broodnest, and nectar is coming in, it might get stored in the broodnest, in cells where bees are emerging before she can lay in all of them.

If the brood pattern is spotty and the above do not seem to apply, you may have a poorly mated or inbred queen. You need to either requeen or simply remove the queen and let the bees raise a new one (make sure that the bees have eggs or young larvae to work with).

If you see a spotty brood pattern that is mostly drones (large bullet-shaped cappings): This is a good indication of a laying worker. There are several ways to try and save such a colony, but in the end, it probably takes the least resources to shake the bees out away from the colony (spreading the foragers to other colonies), and to make a new split from a strong colony to replace the laying worker colony. If you place the split where the laying worker colony was situated, it will have the advantage of the returning foragers.

If you find queen cells (not cups, but see eggs or larvae in the cups or see capped queen cells): You should try to determine if they are swarm cells or supersedure cells. If the bees have plenty of room, and they start to make queen cells within several weeks of starting the hive, they are probably supersedure cells, and it means that the bees, for one reason or another, think that the current queen is unsuitable and needs replacing. We always recommend letting the bees supersede when they want to.

If there are several frames of brood and the hive is strong, you can split off the laying queen to form a nuc. Minimally, you'll need a frame of mostly capped brood (and its adhering bees) and a frame of honey. The queen will continue to lay and be in reserve while the new queens in the parent hive fight to the death, and the survivor successfully completes her mating flights.

Many things can go wrong before the new queen starts to lay, and it's nice to have the old queen in reserve. If the new queen doesn't make it or proves to be a poor layer, you can recombine the nuc with the old queen into the hive. You can make a similar kind of split if you find several frames with queen cells on them. Place each frame into a separate nuc, let each of the queens mate, and you'll have established several new—though small—colonies.

If the queen cells are found in a very populous hive with little room for the queen to lay, and especially when they are found on the outside of the broodnest and on the bottoms of frames, they are more likely swarm cells. At this point, the hive has already decided to swarm.

Bee Smart

Many beekeeping resources instruct you to destroy queen cells if you find them. Don't listen to them.

All too often the queen has either already left with a swarm, or is about to do so. Cutting queen cells out of such hives leaves them hopelessly queenless, meaning that they have no resources from which to raise a new queen. If you find queen cells, and you think the hive will swarm, put the old queen (and no queen cells) with a few frames of bees and stores in a nuc or 10-frame box. Leave the queen cells in the original hive and open up the broodnest. The old queen is unlikely to swarm from a small nuc with limited population, and the new queens get a chance to mate. This gives you insurance against mating problems because you can always recombine the nuc with the parent hive.

It won't often be clear if queen cells are for swarming or for supersedure. You will have to use your judgment. If you see 20 queen cells in the hive, they are probably swarm cells—anything less is more ambiguous.

If your bees are lacking stores (no capped honey, no pollen): They are likely living hand to mouth, consuming food as it comes in. This is not necessarily a bad thing, as it's how bees have survived for 100 million years, but the situation warrants some attention.

If the bees are consistently raising brood—you see brood in all stages, from egg to capped—then there really isn't anything to worry about.

If the bees have shut down brood rearing—you don't see any eggs or open brood—it means there isn't enough food coming in to raise brood. Again, this isn't anything to be concerned about. Food-motivated breaks in the brood cycle are part of the bees' normal functioning and help bees manage diseases and pests. However, if this goes on for more than a couple of weeks, you should consider feeding the bees, especially in a first-year hive that you are trying to have build up comb.

Bee Bonus

One of the most astute observers of honeybee behavior in history was blind. François Huber, with the help of his wife and his servant, made many of the important discoveries regarding honeybees and their behavior that make modern beekeeping possible.

Once the bees have built up eight frames of brood, they have reached critical mass, will start to expand rapidly, and should be largely self-sufficient. Before this point, you need to keep a close eye on the colony.

Having more than one or two hives gives you the opportunity to compare them to one another. A suffering hive quickly becomes apparent when you see three other hives around it thriving. Observation is the key to knowing what is going on in the hive.

Inspecting the Outside of the Hive

The easiest and least invasive observations happen from outside the hive.

Bees fly in and out of the hive in relation to time of day, time of year, temperature, weather, and so on. If something seems out of whack, especially when compared to other hives in the yard, you should investigate the reason. Here are some things to watch out for:

- If there are few bees foraging when they should be, inspect the hive for disease, brood, stores, and a laying queen.
- If there are too many bees flying in and out of the hive, it might be being robbed by bees from other colonies. Look for fighting at the entrance, and menacing-looking bees flying around with their legs splayed open. Closing off

most of the entrance so that the bees are better able to defend their hive will help. You can also move the hive and keep the entrance restricted.

- If you notice a large number of drones flying in and out of the hive, you should look for queen cells. Chances are the colony is raising a new queen (for either supersedure or swarming).
- If bees are streaming out of a hive like crazy, a swarm might be issuing or the hive may be absconding. You can try to catch the queen as she leaves, or follow the swarm and see if you can capture it. If you get the queen, the swarm will settle wherever you place her. Your best option is to put her inside a hive body with a queen includer.

Bee Smart

Stay alert. Just the other day we had a nuc abscond in our backyard. The queen appeared on the top cover, and we were able to capture her without any difficulty. The rest of the bees followed right behind.

- Bees die every day, and you will often see some old or weakened bees crawling around the ground of the apiary. If you see an excessive number of crawlers, or a pile of dead bees outside a hive, you should investigate further. A pesticide kill could be the reason. Foragers could have brought poison back to the hive, killing many of the inhabitants. A virus could also be the cause, as could nosema or tracheal mite, so look for those signs as well (for more on these diseases, see Chapter 10).
- Disturbed soil in front of the hive can indicate skunk activity.
- Brownish streaking on the outside of the hive can indicate dysentery. Most often you will see this in the spring or after periods of confinement. Dysentery can have several causes (long confinement, nosema, etc), but generally improves with forage being available (or feeding) and ventilation if the hive is damp.

Disease and Parasite Specifics

Diseases and parasites are of concern to any new beekeeper, especially because you have never seen these things before. In all cases, the best prevention and treatment is to keep strong, healthy hives of bees.

Common Bee Diseases

Here is a quick overview of the most common bee diseases:

- **American foulbrood (AFB)** is the most dreaded of honeybee diseases. Many states require the bees be destroyed and the hive burned. Some states might allow you to sterilize equipment, including comb, with gamma radiation, and some allow for the use of antibiotic treatments, which predictably, has led to resistant strains. We recommend you contact a local bee club or state agency for advice as to what is legal and recommended in your area.
- **European foulbrood (EFB)** is much less serious than AFB, and will generally clear up on its own. It is a stress disease, meaning that its causative organisms are present in most hives all the time but only show clinical symptoms in times of stress. EFB is more likely to afflict uncapped brood, while AFB kills most brood after capping.
- **Deformed wing virus (DWV)** causes wings that look tattered and malformed. Data suggests that this virus is also present in most honeybee colonies, and is only expressed under stress. There also is a correlation between DWV and varroa infestation. If you see symptoms of DWV, you are likely to find varroa.
- **Sacbrood virus** (or sacbrood) is yet another virus found in most hives. It isn't uncommon to find some of this in the spring as the bees are first building up, but it generally goes away on its own as forage becomes available. Brood is found dead (often resembling EFB) and the remains are contained in a "sack" of tissue. If it is a persistent problem, try requeening from stock that does not show symptoms.
- **Chalkbrood/Stonebrood** are also common in the spring, and tend to clear up with drier weather and increased forage. Some stock is much more susceptible than others, so requeening is the best solution for persistent problems.
- **Varroa mites** did not appear in this country until the mid-1980s. Their impact on beekeeping has been immeasurably negative, and they are generally credited as the largest single threat to honeybees. We (and other small cell beekeepers) stopped having any kind of varroa problem when we regressed our bees to small cell combs. We see a mite from time to time, but they have become a nonissue.

  Varroa mites look like tiny reddish brown pin heads and can be found on older larvae and adult bees. They feed on the bees themselves, impacting their health and longevity. The wounds they make when feeding on the bees provide access to other parasites and diseases.

- **Tracheal mites** are virtually indistinguishable from *Acarapis externis* and *Acarapis dorsalis* mites, which live (apparently harmlessly) on the body of the honeybee. Tracheal mites never seemed to exist before the upsizing of the honeybee, and it's likely that the larger bee created an opportunity for an external mite to move inside. The good news is that resistance to tracheal mites is easy to breed for, and resistant stock is widely available in both commercial and feral populations.

  Tracheal mites get into the breathing tubes of the bees and breed there, interfering with their ability to breathe and ultimately killing them.

- **Nosema** is another very widespread disease, caused by a type of fungi called microsporidian. Only when the infection reaches high levels in the gut of the honeybee does it cause a problem. There are two well-known types of nosema in honeybees: nosema apis and nosema ceranae. Both have slightly different symptoms, but are easy to confuse under the microscope. It was thought until recently that nosema ceranae was fairly new to the United States, but new analysis of old samples taken by Tony Jadczak, the Maine state bee inspector, shows a 30 percent infection rate in 1985! Feeding a little can often help with nosema, unless the case is too far gone, in which case the bees are likely to have trouble taking the feed. A telltale sign of nosema ceranae is when bees that are low on stores won't take feed.

- **Wax moths** deserve at least a brief mention. Brood comb and comb with pollen that is stored away from the bees will become a target for wax moths, and they will eventually eat away the comb and dig into the wood. Extended freezing temperatures will kill larvae and eggs, so unused equipment can be stored outside or in an unheated storage shed through the winter if it is cold enough where you live. Otherwise you can freeze the frames in a chest freezer for three days and then store them in sealed plastic bags. Wax moths don't like the light, so unused equipment should either be stored in sealed stacks after freezing, or the boxes should be offset diagonally so that light can easily penetrate. The best way to keep wax moths out of your equipment is to keep bees in all equipment.

- **Small hive beetles (SHB)** should be treated much like wax moths. The adults are dark brown/black oblong beetles about ¼-inch around. Be on the lookout for the slimy larvae in less occupied areas of the hive. Keep the hives strong, keep equipment occupied, or freeze, seal, and store it. SHB larvae can take over a hive quickly. Some trap designs work well for helping to deal with SHB short term while a hive gains its strength.

- **Paralysis virus** should be suspected if you notice "greasy" looking bees inside and outside the hive. The virus causes the bees' hairs to fall off, and leaves the shiny exoskeleton (looking more like a wasp than a bee). You might also see quivering bees on the bottom board or out in front of the hive.
- **Colony collapse disorder (CCD)** is not a disease, but a set of symptoms with no established cause(s). There is a lot of media attention and hype in connection with CCD, but the basic facts have been largely ignored. Commercially managed honeybees (with a few exceptions) are treated like other agricultural livestock: high population densities, artificial feeds and treatments to increase yields, and breeding stock selection geared toward commercial exploitation rather than the self sufficiency of the bee. In short, more yield is expected from the same resources year after year. Such systems eventually break down for one reason or another. The overall cause for CCD likely has to do with the interactions between bees and microbes, and any research performed in this area that is not on bees kept without treatments and artificial feeds will never reveal the true nature and importance of these relationships.

  It's important to note that bee die offs have been noted throughout recorded history. We have a copy of a report from the Secretary of Agriculture in 1869, describing the disappearance and disease of bees in 1868. Apart from the slightly antiquated language, the paper reads like a report on modern day CCD.

The following sections describe specific symptoms to look for, what they might indicate, and what might be done.

Perforated Cappings: Chilled Brood, American Foulbrood

Small perforations in the cappings generally indicate dead brood in the cells. The perforations are from gas pressure building up inside the capped cell as the dead brood rots. Perforated cappings can be caused by chilled brood, which happens when adult bees can't keep the brood warm. Either a cold snap caused the bees to cluster and abandon some of the brood on the outside of the nest, or a loss of adult population left the bees unable to keep all the brood warm.

The loss of capped brood represents roughly the same investment in food, nursing, and resources as a newly emerged adult honeybee. Losing any capped brood represents wasted resources.

Of more concern than chilled brood is American foulbrood (AFB), which can also have perforated cappings. AFB is a spore-forming bacterial infection that can wreak havoc in an apiary. It is the most important disease to keep an eye out for, and hopefully the one you will find the least often. If you see perforated cappings, use a matchstick, a stiff piece of straw, or something similar to probe and mix up the contents of several symptomatic cells. You are looking for a rope-like texture as you pull out the probe. A sticky mass that stretches out and snaps back into the cell like a rubber band is a good indication of AFB, whereas a liquid-y mixture or a dried-out mummy is not. If at all in doubt (or curious), you can send a sample to the USDA for free disease analysis.

AFB has a distinctive *scale* that dries in the bottom crease of the cell and is difficult to remove. However, not all debris in a cell is scale, and not all scale is due to AFB.

A larger hole in the capping indicates that the bees are chewing open the cappings of brood with disease or parasites. Seeing a few of these cells in the hive is a good sign, as it indicates that the bees can sense a problem, and have a way to deal with it. On the other hand, if there are many such opened cells it might indicate a bigger disease problem that the bees are trying to deal with.

def•i•ni•tion

Scale is the mass left behind by decayed brood.

Discolored Brood: European Foulbrood, Sacbrood, Chalkbrood, and Stonebrood

Open brood that is still coiled in the cell but has turned tan, brown, yellow, or any color other than pearly white means the brood is dead or dying. A number of causes are possible. Chilled brood will look like this, as will European foulbrood, and sometimes sacbrood. If you see dead, open brood, you should try to determine the source. Dead brood that are lying lengthwise in open cells are usually darkened in color. A larval tongue that is left touching the top of the cell is a good indicator for AFB, and should be tested for ropey-ness and perhaps samples sent to a lab.

If, when probed, the remains seem to be contained in a pouch or sack, it is a good bet that it is sacbrood *virus*, which is present in most colonies, and usually only shows up in times of stress.

If the larva are twisted in the cell (rather than lying tightly coiled in the cell), EFB should be suspected.

If you see hard, solid pellets inside the cells (ranging from white to brown), you likely have a case of chalkbrood or stonebrood. These remains are often referred to as "mummies," and can sometimes be found on the ground outside the hive as well. Chalkbrood should clear itself up, or the hive requeened with more resistant stock.

Disfigured Wings: Tracheal Mites and Nosema

Bees' two sets of wings are generally attached to one another by hooks, making it appear as if they have only one set. When one or both sides become unhooked, the wings separate. When this happens only to one side, it resembles the letter K, hence the term *K-Wing*. If you have just a few bees with K-wing, there is no cause for concern. Many bees with K-wing indicate something wrong, such as tracheal mites or nosema.

An obvious case of K-wing. This bee also appears to be suffering from viruses (deformed wing and paralysis).

Greasy Bees: Paralysis Virus

Greasy-looking bees inside or outside the hive often indicate a paralysis virus of one kind or another. The slick look is because their hairs have fallen out as a result of the virus.

The fuzzy, healthy bees are pushing this diseased, almost hairless bee with paralysis virus out of the hive.

Dying Emerging Bees: Lack of Stores and Nosema

If you see dead emerging bees with their heads sticking out of the cell and their tongues out, they usually starved to death. There was simply not enough food to pack in the cell before capping, and/or the bees were not fed while emerging. Look for a lack of stores. If starving bees are in this position and will not take feed, suspect nosema.

Common Comb Issues

Not every frame of comb that the bees draw is perfect by beekeepers' standards.

Misdrawn Comb

In a managed hive, straight comb is interchangeable with other straight comb anywhere else in the hive. The presence of straight comb causes bees to draw more straight comb. Beekeepers like straight comb because it can be harvested and extracted easily. It can also be inspected for disease. In a natural colony (without frames or comb guides), the comb might twist and turn, making it difficult to harvest and extract as well as inspect for diseases.

Bee Smart

Never cull comb from your hives without first harvesting honey from it if at all possible. It takes a lot of honey to make a frame of comb, so don't waste it by culling it without first getting it filled with honey.

Bees will sometimes draw two layers of comb on a frame, or if you're using foundation, they sometimes draw comb off the foundation. This misdrawn comb should be culled. Usually, when we see such comb, it has brood laid in it that we don't want to kill, so we slowly migrate it to the outside of the broodnest, and then up to a higher box to be filled with honey, or if it is bad enough that it will interfere with other combs, we simply cull it when it's free of brood.

Bees will also sometimes draw honeycomb out extra thick, so it intrudes onto the next frame. You can trim this for a little honey treat or cut off the excess comb when you harvest the honey.

Large Cell Comb

Frames of comb with a high percentage of very large (drone/honey storage) cells are less desirable than combs with a higher percentage of small cells. Bees usually want about 15 percent of their comb to be drone comb, and they usually make it on the outside of the broodnest. If you find frames with more than 15 percent drone comb, or where the drone cells are in the middle of the comb, it's good practice to gradually move these frames to the outside of the box, and then up to a higher box.

Contaminated Comb

Even those of us who don't use pesticides inside our beehives have to face the fact that they are ubiquitous in the environment, and all kinds of pollutants are present in even the most pristine beehives.

Dr. Jerry Bromenshenk, a honeybee researcher in Montana, reports that he hasn't been able to find a honeybee sample free of PCBs in the last 20 years. At some point, the comb contains more toxins than one would like and it should be rotated out. Local conditions such as the amount of pesticide use and environmental pollution should be taken into account, but good comb should last at least 10 years in a treatment-free operation.

The Least You Need to Know

- Most of what you will learn about your bees comes from patient observation.
- Think through your hive inspection before you begin: have your smoker lit, wear appropriate clothing, and don't do anything without a good reason.
- The hive inspection is your chance to compare what you think is going on with what is actually going on. Every inspection is a tremendous learning opportunity.
- Hives are most susceptible to diseases when they are stressed.
- Monitor combs and migrate misdrawn or large cell combs outside the broodnest.

Part 3

Beekeeping the Old-Fashioned Way

Although bees don't use moveable frames or stackable boxes in the wild, we strongly suggest a model and management scheme that is as close to nature as you can get while still keeping the bees under your care. That means putting long-term gains before short-term profits.

You'll learn how to manage your hives using unlimited broodnest, in which the queen is allowed unrestricted access to the entire colony. You'll also find out how bees have been enlarged and why naturally smaller bees are so much healthier. You'll learn what a "drone right" colony is and why it's best for the bees.

Diseases and parasites can show up from time to time, but with rare exceptions, you'll learn the wisdom of not treating the bees with any chemicals or other substances. Instead, you'll find out how you can manage disease using occasional feeding, requeening, equalizing, or perhaps by allowing a period of broodlessness.

Finally, you'll get a crash course in bee sex. Bees raised in your area have the best chance of surviving and thriving without a lot of intervention on your part, so breeding localized bees is really the best way.

Chapter 9

A Hands-Off Approach to Hive Management

In This Chapter

- Embracing unlimited broodnest
- Understanding the upsizing of the honeybee
- Regressing your bees to small cell
- Using foundationless frames, foundation, and Honey Super Cell
- Making your colony drone right

Honeybees are simply amazing. They thrive in the constant summer of the equatorial tropics and overwinter in Alaska where temperatures remain well below freezing for several months. Even in the dead of winter, bees can maintain a temperature of 94 degrees in the center of the broodnest while it's below freezing only inches away.

If you live in an area where regular daytime temperatures fall below 50 degrees and there's a period of time in which no nectar or pollen is available, you must make sure that the bees have enough stores of real honey and pollen to carry them through. In the Northeast, it's generally considered appropriate to leave at least 80 lbs. of stores per hive for the bees to get through winter. This amount should increase the farther north you go. When looked at from the perspective of maintaining a healthy, sustainable bee population, it's better to leave too many stores in the hive than too few. You want to leave enough so that the bees have ample stores to build up on before the spring flow.

Unlimited Broodnest

One of the best ways to ensure your bees have enough room to store honey and to sustain themselves is to adopt a hive management practice called unlimited broodnest.

Unlimited broodnest management allows the bees to form their boodnest into more of a "bell curve" shape up the center of the hive, up into the fourth or even fifth box, and to store honey off to the sides. As the honeyflow comes on full force and brood rearing slows, the bees will place honey in the upper boxes, pushing the queen down into the lower two or three boxes.

With this practice you keep the equivalent of at least three deep hive boxes for the bees and their stores. An unlimited broodnest lets the bees take advantage of extensive stores and build the large populations of foragers needed to collect large amounts of honey.

> **Bee Bonus**
>
> Remember that food stores are like capital to the bees. They cannot raise brood to prepare for an early spring flow without an account full of honey and pollen.

Other less sustainable management strategies use stimulative feeding at specific times and frequent requeening in order to achieve these goals, but for the long term, it is better to leave honey with the bees to accomplish such goals.

Resources Available in House

Abundant stores in a large unlimited broodnest allow the bees to subsist through lean times, to forage for scarce resources, and to build up prior to and in response to flows. This is the key difference between the management practices presented here and what you read in most other books and beekeeping guides. Restricted broodnests requires frequent intervention by the beekeeper. A larger hive with an unrestricted broodnest is more stable.

Unlimited Broodnest vs. Honey Supers

Unlimited broodnest differs from conventional approaches to modern beekeeping. Beekeepers are usually taught to keep only one or two deep boxes for overwintering and to use a queen excluder to prevent the queen from laying anywhere but in these two boxes. This encourages the bees to store most of their honey in the top honey supers, which beekeepers can then remove before winter and harvest the honey from them.

Beekeepers prefer to separate the broodnest from honey stores for a wide variety of reasons, most of them having to do with convenience.

Using conventional honey supers also simplifies the harvesting process, as whole boxes of honey can be removed at once. Treatments that can contaminate honey are applied only to the broodnest and not to the combs that will be harvested.

When beekeepers treat the broodnest, they cannot harvest any honey stored in that area. In contrast, unlimited broodnest management uses no queen excluder, and because no treatments are used, comb in the broodnest can be used for honey storage, and the hive bodies are interchangeable.

While beekeepers who use treatments are prohibited from using them in honey supers or during honeyflows, bees do not discriminate between treated and untreated areas of the hive.

Bee Aware

"Having an untreated area of a treated hive is like having a no-peeing section in a swimming pool."

—Michael Bush, beekeeper

How to Achieve an Unlimited Broodnest

In order to achieve an unlimited broodnest, you should keep the broodnest open by occasionally moving good drawn worker comb into the broodnest area and relocating capped honey frames toward the outside of the boxes.

You should harvest honey only from the fourth box and above, leaving what the bees put in the lower three boxes for their own use. Leaving these extensive stores for the bees allows them to maintain themselves through dearth.

Comb-Waxing Poetic

The importance and significance of the wax comb produced by the honeybee cannot be overstated. Without comb, honeybees cannot exist.

Roles of Comb

Comb is much more than just a pantry; it is the skeleton of the hive superorganism. It is also womb, nursery, food cellar, physical support, airflow baffling, thermal conductor, communication conduit, physical structure and liver of the hive.

Bee Bonus

Honeycomb structure is an ideal use of material for both strength and storage capacity. You'd be lucky to find 1 pound of wax in four full deep frames of comb, but that single pound (as comb) can store 26 pounds of honey!

Wax tends to absorb and hold contaminants and toxins (no matter where they come from). This ability for the wax to absorb and encapsulate contaminants reduces the bees' exposure over the short term, but comb that is heavily contaminated can be detrimental to the bees and can contaminate products of the hive.

The bees' ability to properly maintain their nest over a long period is integral to their well being.

Construction

Bees instinctually build (draw) comb using wax flakes secreted by their bodies.

When forager bees return to the hive with a full crop, or honey stomach, they pass the nectar to house bees, who are younger and at the age where they produce wax the best. The house bees then deposit the nectar into cells.

If all the cells in the comb are full and the forager bees have nowhere to deposit nectar, young workers will involuntarily produce wax profusely. They will draw comb anywhere there is space.

This is an important aspect of bee behavior. The house bees will store incoming nectar if they can. If there is nowhere available to store it, the holding of the nectar in their honey stomachs will trigger the impulse for their wax glands to begin producing wax flakes. Conversely, when the flow is over, the bees will not use food that can be stored for future use to produce wax for comb that is not immediately needed.

When constructing comb, the bees *festoon* like a barrel of monkeys with interlocking legs and arms. Bees hand their wax scales up the chain where a few bees at the end of the line use their mandibles (mouth parts) to shape the wax into perfect hexagons.

Bees build honeycomb (as opposed to brood comb) only when more room is needed to store incoming nectar. If there is no space to expand, or if the space available isn't desirable for some reason, the bees will store nectar in the broodnest as bees emerge, and probably swarm.

def•i•ni•tion

Festooning is when the bees hook their legs together and hang freely in a chain, forming a hanging cluster of bees. Festooning allows the bees to pass wax flakes from bee to bee and to direct heat to where they are drawing comb, making the wax more malleable.

The Size of a Cell

Brood comb comes in three distinct sizes as well as some odd-size transitional cells placed between cells of differing sizes. Worker cells are the smallest, followed by drone cells. The largest are queen cells. Within the space of a linear inch you can fit five worker cells, four drone cells, or three queen cells.

Cell Size Matters

What is a proper-size cell? What do bees build in nature? How does that compare with what beekeepers are using in the field? These are important questions, ones that the modern beekeeping world didn't ask until the late 1980s when Ed and Dee Lusby's observations led the Tucson Bee Lab to investigate. The result was the conclusion kept honeybees are unnaturally large.

Foundation's Foundations

Before the invention of embossed wax *foundation*, bees built whatever size comb and individual cells they cared to. A century ago, the size of a worker cell was at most 5.08 mm and probably smaller in the middle of the broodnest. The size of the cell limits the amount of food that is fed to the growing larvae, and also limits the size of the emerging bee.

Foundation revolutionized beekeeping by making forcing bees to produce straight, interchangeable combs.

def•i•ni•tion

Foundation is a sheet of beeswax embossed with hexagonal comb pattern. Beekeepers insert foundation in the hive frames, and bees use it as their template for creating comb.

Bee Bonus

The size referred to as *small cell* is 4.9 mm. Currently, standard foundation in the United States is 5.4-mm. The difference in volume between 4.9-mm and 5.4-mm cells is around 30 percent.

One of the first functions of foundation in the beekeeping industry was to enlarge the honeybee with a "bigger is better" philosophy, and an attempt to make the bee's tongue long enough to gather nectar from purple clover.

Beekeepers and foundation manufacturers increased the size of the cell imprint in foundation to 5.4-mm, and the bees followed suit, making larger cells. The bigger cells meant that nurse bees put more food in the cell and the larva grew larger. Surprisingly, it worked—we started getting larger bees!

What's most interesting about this process of artificially (and not genetically) enlarging the bee is that, to a certain extent, cell size is heritable. It takes many generations after rebuilding the broodnest from scratch for the bees to regress back to a natural size on their own.

The Downside of Large Cells

While the initial impulse to enlarge the honeybee was for a longer tongue and, ultimately, an increase in honey production, the opposite may actually have happened. Small cell comb provides a more dense population of bees per square inch, giving the hive a bigger, more productive work force. According to Dee Lusby's data, small cell bees can actually outperform large cell bees in honey production.

Large cell worker cells are close in size to naturally sized drone cells. Remember that diseases and pests commonly show up in drone brood before worker brood, as the larger size larvae (and their food) provides more resources for pests and disease. When unnaturally sized worker brood mimics naturally sized drone brood, there is a greater chance of infestation in the worker brood.

Varroa mite reproduction occurs in the developing bee. There is evidence that the gestation period of large cell bees is about one day longer than that of small cell bees, providing the varroa mite with an opportunity to reproduce in greater numbers.

Although one day may seem insignificant, the timeline for varroa to mate and reproduce is tight, and these few hours reduces mite reproduction by half (or better).

Although varroa is the best reason to regress bees to a natural size, it is far from the only one. Large cell bees are not proportioned properly with regards to weight versus length. Their extra mass (and aerodynamic drag) benefit from longer wings, but not from more massive flight muscles. An apt analogy would be to enlarge an airplane by 30 percent in all aspects (cargo capacity, wingspan, length), yet leave the engine as is. You could not run an airline with such planes profitably.

The artificial enlargement of the honeybee puts it out of sync with the ecological systems it has co-evolved with over millions of years. We may never realize all of the implications of such a seemingly subtle shift, but fortunately, with time and patience, it is something that can be reversed.

The Upside of Small Cells

Small cell beekeepers report many differences in their bees when compared to large cell bees.

Small cell bees …

- Are better at flying and landing.
- Forage earlier in the morning and later in the evening.
- Forage in colder weather and hotter weather.
- Forage over a wider area on more diverse flowers.
- Are more resistant to diseases and pests.
- Are better at ridding the hive of diseases and pests.

To put it simply, small cell bees are better tuned to the environment in which they evolved.

At first, using small cell bees seems like a common "cure all" claim that is too good to be true. But keep in mind that we are talking about comparing a population that is a natural weight with one that is 30 percent overweight. In humans, a doctor

Bee Aware

The size of bees changes naturally with latitude and altitude. This shows that size is one way in which bees can adapt to their local environment. It is also one way in which they have been thrown out of balance.

would recommend losing weight to anyone 30 percent overweight. People who are 30 percent overweight are at higher risk for a number of health problems, including cardiovascular disease, diabetes, depression, and cancer.

For many beekeepers, reducing the cell size back to something resembling its natural size has been the key to maintaining a healthy hive without treating it at all.

Regressing to Small Cell Bees

Regression is the process of converting unnaturally oversized bees to small cell. It can take as little as two months or many years to accomplish, depending on the technique you adopt, the bees you start with, and your attention to the process.

Regressing your bees to small cell can be challenging, and it's not something to be undertaken lightly. It will likely be difficult, and will set your bees back a few weeks at best. However, this short-term investment will pay off in the long run if you are not going to use treatments.

If you purchase bees that are already regressed, then you do not have to regress them.

Bee Bonus

You will never know in advance, but some package bee suppliers have some of their bees on a plastic combination frame/foundation made by Mann Lake. Mann Lake PF-100 and PF-120 products are sold as standard (which is 5.4mm), but in fact have a cell size of 4.95mm (nearly small cell, which is 4.9mm). Package producers often have a wide range of equipment in the field and might even source bees from other beekeepers, so just because one package comes small from a particular supplier doesn't mean the next one will. In short, never count on this being the case. Be prepared for package bees to be on large cell and require regressing.

Why Regress?

Our experience—and the experiences of many other beekeepers—indicates that regression is necessary. We certainly have some bees die during the regression process, but they don't all die, and once regressed they don't succumb to mites. Most beekeepers are focused on varroa mites as the biggest problem in beekeeping. We don't see any appreciable level of mites in our regressed hives, and our state inspector doesn't see them when he performs his annual inspection.

HSC

The quickest and most reliable method of regression is to use Honey Super Cell (HSC). With HSC, your regression will go from a risky multi-year project to a two-month process.

HSC is an injection-molded plastic comb with 4.9mm cells. HSC is not foundation, it is fully formed plastic comb. HSC is not a perfect replacement for wax comb. The bottoms of the cells are flat and the cell walls are thick so that the density of brood is closer to that of 5.4mm comb (the cells are 4.9mm across, but 5.4mm center to center). HSC only comes in deep frames. If you want to run medium boxes, you have to trim the frames using a table saw.

Understandably, bees are initially reluctant to use a plastic comb. When getting the bees to start rearing brood in HSC (and thus regressing them), keep a few points in mind:

- HSC regression is most easily accomplished with package bees (see Chapter 4), as they don't come with any comb. Bees prefer wax to plastic and will be reluctant to use the HSC if other comb is available. Regression with HSC is possible with nucs or existing hives, but it's more difficult.
- Because bees will almost always choose to build wax comb over using new HSC, it's important to give the bees no choice. No other comb or empty space for the bees to build comb should be available.
- Rubbing the HSC frames with beeswax makes the plastic more appealing to bees. If you don't have access to wax, either your own or from another beekeeper, purchased beeswax will do. It's not necessary to coat every bit of the cell faces, but you want to have enough wax to give the combs a homey smell.
- Adding honey syrup or sugar syrup (one part sugar dissolved in one part water) to the cells makes them more palatable to the bees. Use a ladle or watering can to pour the mixture into the cells.

Bee Aware

If you have access to honey that you are sure is from untreated, healthy bees, you can use honey syrup instead of sugar syrup for your HSC regression. However, sugar is preferable to honey that could be harboring disease or antibiotics. While a strong, established colony can often deal with diseases, packages are vulnerable. It isn't worth the risk for the small amount of sugar that the bees will ingest.

Use a full 10-frame box of HSC for each package of bees. This will give the bees plenty of space to work during the two months it will take for a complete regression. Once you have the HSC combs prepared, you can install the bees as explained in Chapter 7. Feed a package in a 10-frame box until they have four frames of capped stores.

When you install bees on new HSC, be sure to use a queen includer to keep the bees from absconding. Once the bees have a frame of capped brood, they will be committed to the hive and you can remove the includer.

Even though you have given the queen no other options, she will probably be reluctant to lay in the HSC and it may take a few weeks before you see any sign of brood. Be patient. Eventually she will give in and lay. Meanwhile, the workers will have begun to outline the cells with their own wax and fill them with pollen and nectar.

Bee Smart

The reason we advocate this regression method above all others is time. Although you lose a few weeks at the beginning because the bees are reluctant to use the HSC, within two months, all of your bees will be small (as all bees from the package except for the queen will have lived out their lives, and only bees raised in the HSC will be present). As you need to expand or split, you can use either foundationless frames or small cell foundation. Once all of your bees are small cell, they will remain that way without having to rear the brood in HSC.

As the broodnest expands, follow the techniques for adding boxes and pyramiding up that we discussed in Chapter 7. Once the HSC has been used by the bees, future bees will accept it freely. Just remember that while HSC will always produce small cell bees, you will never have the density of bees per square inch that is possible with small cell foundation or small cell foundationless comb.

Small Cell (SC) Foundation

Worker foundation can be found in three different sizes:

- 5.4mm (the most popular, referred to as large cell or LC)
- 5.1mm (generally an intermediate size for regressing)
- 4.9mm (small cell, or SC)

Most commercially available bees are raised on large cell (LC) comb, and it is often difficult for them to draw out small cell (SC) comb because their bodies are too large.

Another approach to regressing bees is to simply use SC foundation, let the first results be imperfect, and rotate out the bad comb as the cells get smaller. This method works with packages, nucs, or even full-size colonies with the caveat that the bees remain vulnerable to disease and pests until they are regressed below 5.0mm, which might take a season or two. It is a roll of the dice as to whether the bees can survive long enough to get to this point.

Large cell bees will, however, be able to draw out intermediate 5.1mm comb fairly easily, and once regressed onto that size, they will be able to draw small cell comb. This is a bit more of an orderly process, and is useful not only for packages, but also is a way to regress a new nuc or even a fully established colony. Simply feed the 5.1mm foundation into the broodnest, and once most of the brood is emerging on the 5.1mm comb, start with the 4.9mm foundation. Depending on the climate/weather and strength and size of the colony, this can take from one to three years.

A slightly faster version for a large, established colony would be to remove all but the brood comb from the colony, and add a full box of 5.1mm foundation. A strong, crowded colony on a flow should be able to draw out and use the foundation quickly. When the broodnest is mostly 5.1mm comb, repeat the process with the 4.9mm foundation.

Bee Aware

Bees will build comb when resources are available. Ideally, regression should commence in the spring, when forage is plentiful. Pay attention to the weather and climate in your area and talk to local beekeepers. Don't expect the bees to draw comb during a dearth or too late in the season.

Foundationless Frames

Some people have had luck simply putting their package bees onto foundationless frames, rotating empty frames in, replacing those with larger cells, and letting the bees regress themselves. Doubtless they will do this eventually, but our experience is that this is too slow a process to get them regressed before pests and disease become a problem. However, if you are able to get package bees that were raised on the small Mann Lake plastic comb, a foundationless approach will work fine.

Getting Good Comb Drawn

Drawn comb represents a huge input of resources by the bees and the beekeeper. The sooner you can get properly sized combs drawn, the more time and energy you and your bees will save.

As noted earlier in this chapter, bees draw a variety of cell sizes, from the smallest worker cells to the larger drone and nectar/honey storage cells. Getting your bees to draw the right size cells in the right place at the right time is fundamental to the success of the hive.

Worker comb is the most labor intensive for the bees to produce as the comb has the most cells per square inch and uses the most wax per square inch. Drone comb and honeycomb require fewer resources to draw as fewer cells are needed to fill a frame and less wax overall is used.

There are five factors involved in drawing comb: need, heat, space, number of bees, and incoming nectar. These factors are all interrelated and interdependent on each other.

For instance, incoming nectar might mean that there is a need for honeycomb to be drawn, unless empty storage already exists. If the bees are cold, they may use the incoming nectar to generate heat, rather than store it for future use. Once the bees have enough energy to generate surplus heat, they can begin to draw comb, as the wax requires heat to make it malleable. A critical mass of bees is needed to generate this surplus heat.

Anticipation of a honeyflow may mean that the bees draw out worker comb so that the queen can have room to lay new eggs. It may also mean that the bees begin to draw honeycomb to prepare for future stores. A honeyflow without sufficient space to build honeycomb may result in the storage of nectar in the brood comb, plugging up the broodnest and leaving the queen without room to lay more eggs.

It is up to the beekeeper to make sure that the bees have the space they need to draw new comb where and when they need it.

Worker Comb

Properly drawn worker comb is crucial for maintaining the small cell size of your regressed bees. The easiest way to maintain small cell bees is to provide them with small cell foundation in the core of the broodnest. The foundation gives the bees a template to work from.

Unless you are milling your own foundation from known clean wax or have access to foundation from other beekeepers that you know is free from beekeeping chemicals, you may be exposing your bees to potential contaminants in commercial foundation. To minimize contamination, use the commercial small cell foundation in the center five or so frames of the hive body and, as the broodnest expands, insert foundationless frames. Keep an eye on the comb. If your bees still need help drawing small cells in the core of the broodnest, feed in small cell foundation. Otherwise, keep feeding in foundationless frames. Some bees will be more skillful at drawing comb than others.

Bee Bonus

Foundation also gives the queen quicker access to egg laying, as she can deposit her eggs on the surface of the foundation while the cells are in the process of being drawn out by the workers. Foundationless comb is drawn down (as opposed to drawn out) by the bees and does not give the queen this early access.

Drone Comb

In a natural system, bees typically devote 10 to 15 percent of their comb to drone comb, and they like to build it on the outskirts of the broodnest.

Conventional beekeeping practices advocate eliminating drone comb from the broodnest area, which forces bees to build drone comb between boxes, in the space between the bottom of one frame and the top of another.

The practice of using all worker foundation also prevents bees from creating a *drone right* colony and forces them to put drone comb and brood where they can.

Beekeepers justify removing or preventing the construction of drone comb from around the broodnest because they consider drones to be a drain on the hive's resources and because varroa mites tend to concentrate on the drone population. When the bees build drone comb between frames rather than in frames around the broodnest, it's easier for the beekeeper to check the drone comb for mites. However, when drone comb is placed between frames, cells are opened when the boxes are separated, resulting in the death of many drone larvae. Since the beekeeper doesn't want drones (or he would have provided a place for them), the death of the drones is perceived as a good thing, or at least not a big deal.

def•i•ni•tion

A **drone right** colony is one that has enough drone comb or room for bees to draw drone comb in the area of the hive they want it. A drone right colony will not desperately draw drone cells in any available space.

But such beekeepers are failing to take into account the investment in energy the bees have already put into a drone larva that is old enough to check for mites. The bees' compulsion to have 10 to 15 percent of their population be drones is so strong that as soon as the developing drone brood or comb is destroyed or removed, the bees will attempt to replace it. This is a tremendous waste of energy in the hive, as the bees will make this same investment over and over, whereas they would stop and move on to other priorities if the drones were allowed to mature.

Another place that drones show up when suppressed is in the honey storage areas. If all worker foundation is used, and all comb in the broodnest is tough with age and built-up cocoons, it is difficult for the bees to enlarge the cells in order to lay drones. In honey storage areas, however, the wax is supple, and in most cases contains no cocoons. These cells can be easily reworked into drones, exactly where the beekeeper doesn't want them.

Bees should be allowed to structure the broodnest as they deem necessary. If you use worker foundation, cutting an inch off the bottom of the foundation is a good way to provide enough space for the colony to be drone right, as are foundationless frames.

Make sure to leave your bees the resources to build the drones they want. If the hive is creating more than 15 percent drones, you can begin the process of moving the drone comb to the outsides of the hive body and ultimately up into the honey storage areas.

The Least You Need to Know

- Unlimited broodnest gives bees working capital.
- Unnaturally large bees are more prone to diseases and mites.
- Small cell restores bees to a natural state.
- Honey Super Cell is the best way to regress your bees.
- A drone right colony avoids wasting resources.

Chapter 10

The Yin and Yang of Disease

In This Chapter

- The importance of microbial balance
- Common treatment practices
- How not to treat
- Taking the long view

If you spend time with people who make their living in any kind of agricultural pursuit, you will hear about two concerns above all others—the weather and disease. We can't do anything about the weather (except schedule around it the best we can), but disease presents opportunities—specifically, opportunities to prevent, treat, and most importantly learn about the conditions and organisms that cause disease. What we end up learning is that disease is necessary for health in the same way that death is necessary for life.

The struggle against disease is a necessary part of beekeeping. Honeybee diseases help us make sense of what it means to have a healthy colony. A better understanding of bee diseases offers beekeepers a fuller sense of the complex interactions between bees and their microbial "colleagues" in the hive.

Unfortunately, these lessons have real consequences. Disease can lead to decreased production, dead bees, and, worst of all, the spread of disease. There is more at stake than some intellectual exercise; we need to consider the economic, agricultural, and emotional impacts of disease as well. Most importantly, we need to stay focused on the long-term impacts of our approach to handling diseases.

Balancing Act

Most honeybee diseases (American foulbrood, chalkbrood, stonebrood, nosema, European foulbrood) have antagonistic relationships with one another. In a healthy, natural system, these and countless other unrecognized microbial collaborators struggle both with and against one another, ultimately supporting the bees.

For 100 million years, the combative microbial environment in the hive has been selected for one thing only: cultures that can support bees that thrive and reproduce.

Certainly new and exotic challenges come along from time to time. Some of these are devastating to populations. However, we must keep in mind that disease causes imbalance, which exerts pressure toward balance. This pressure causes genetic bottlenecks, which are scary close up ("the bees are dying"), but ultimately lead to robust populations (unsuitable genetics are culled).

Modern Treatment Practices

The beekeeping industry (including the hobbyist component) has unfortunately sacrificed long-term sustainable for short-term success. Let's be clear here. Virtually all beekeepers (including the old man at the end of the dirt road at the edge of town with a handmade "local honey" sign) use treatments in their hives.

A variety of treatments are used routinely in beekeeping, which range from "synthetic and scary-sounding" to "natural, holistic, and nontoxic to humans." Let's consider them by rough classifications.

Off-Label

It is illegal to use anything in your hive to treat for disease that is not specifically approved for such use. In addition, any treatments that are approved must be used as directed on the label. With that said, if you read any of the online forums where beekeepers congregate, you will read about all manner of unapproved treatments, from essential oils fed in sugar syrup to onion juice, to shop towels soaked in cocktails of toxic synthetic chemicals. The risks of using such treatments—to the bees and to humans that consume their products—are unknown. One cannot even begin to weigh pros and cons if we have no way to know either.

Synthetics

Coumaphos (an organophosphate; essentially nerve gas) and fluvalinate (a pyrethroid; a synthetic version of a chrysanthemum toxin) are the two most common in-hive treatments to fight varroa mites. Both of these treatments harm brood, affect queen longevity and fertility, and affect sperm production in drones.

Bee Aware

Coumaphos and fluvalinate have been used so much by so many beekeepers that any beeswax produced in the United States—including foundation—is almost assured to be contaminated with these chemicals.

Antibiotics

Terramycin (Oxytetracycline) is used in many beekeeping operations to prevent an outbreak of American foulbrood (AFB). AFB is caused by a spore-forming bacteria, and the spores are very persistent—they can remain viable for 50 years or longer. An active infection can be controlled with treatment but will inevitably return as the application wears off and the spores germinate.

Terramycin is a broad-spectrum antibiotic, which means that it kills a large number of bacterial species. One of the side effects is that bacteria keep yeast populations in check, and they inevitably rise when antibiotics are used (this is true in humans as well as in honeybee colonies). There is also some evidence that bacterial species that help keep AFB in check are killed as well.

Fumidil (another antibiotic) is used to prevent and to fight nosema (both apis and ceranae). It's worth noting that the use of fumidil in hives is legal only in the United States and Canada. Fumidil is produced by culturing the causative agent for stonebrood. This is to say that the treatment exploits the antagonistic relationship between microbes that keep one another in balance in the hive.

Organic Acids

Formic acid and oxalic acid (which is unapproved in the United States) are both considered "soft treatments" by the beekeeping community. Many beekeepers believe that using these substances instead of the synthetics to fight varroa is a step forward.

> **Bee Bonus**
>
> Most organic honey production standards around the world allow for the application of organic acids to hives.

It's true that we are not terribly concerned about either of these substances ending up in honey that we eat; both are present in small concentrations in honey naturally. What is alarming is the effect that such substances have on the microbial culture in the hive. In the amounts used, the environment in the hive is greatly altered, and the acids kill some microbes and encourage others.

Natural Treatments

These include essential oils of all types, dusting the bees with powdered sugar, using leaves of various plants in the smoker, and other types of "herbal" treatments. Just because they are naturally occurring substances doesn't mean that they are necessarily good. Consider, for instance, using the oil of the poison ivy plant as a massage oil. Furthermore, these treatments never naturally occur in the concentrations found in these extractions. Essential oils impact microbes (as well as bees) negatively. They are volatile and the smells they release interfere with the natural chemical communication in the hive.

What's Wrong with These Treatments

All of these treatments are harmful to the bees for two important reasons.

First, they severely disrupt the microbial culture that has developed over the last 100 million years or so. Using any of these treatments imposes an environment on them that they have not evolved to handle. The intricate dance between the bees and these thousands of species of microbes that support them all must be allowed to continue without such interference. Otherwise, we will always be chasing the balance that nature alone can maintain.

Second, by propping up genetic stock that cannot survive the conditions in which it lives, we doom future generations of bees (and beekeepers) to an endless march on the treatment treadmill. We must cull unsuitable stock as ruthlessly as nature does, and not indulge its demands like a spoiled child.

Natural Systems and Treatments

With the exception of American foulbrood (AFB, which, if not handled appropriately, can spread quickly and kill both your bees and your neighbors' bees), we don't feel that any action should be taken for any disease beyond making sure the bees have food, ventilation, the appropriate amount of space, and the possibility for requeening. This flies in the face of our instinctual need to "do something" to correct problems. But if you consider that paralysis virus, deformed wing virus, sacbrood, stonebrood, and chalkbrood have no known effective treatments, it only leaves mites, nosema, small hive beetle, wax moths, and European foulbrood we are advocating not treating.

Bee Bonus

We have not had a mite problem since regressing to small cell. Small hive beetles aren't much of a problem in our area, and we only see an occasional beetle. Wax moth is mostly an issue with very weak hives and stored comb, and we try to keep our hives strong and our comb on the hives. European foulbrood, viruses, and chalkbrood occasionally show up but disappear again just as quickly.

As time passes, diseases and parasites ebb and flow, environmental and weather-related cycles and trends play themselves out, and populations evolve. Microbial cultures live out billions of generations through all kinds of conditions, continually adapting toward an unreachable balance between short-term expansion and long-term sustainability. These processes are to be embraced. Fighting them to suit our wants is futile and, ultimately, harmful to the bees and to us.

The Treatment-Free Approach

When people ask us how we run a breeding program for bees that doesn't require treatments, we often respond by saying something like "well, we keep a bunch of bees, don't use treatments on them, and we don't breed from the ones that die." Although such a response is somewhat flippant, it also happens to be completely true.

The reality is that there are so many variables involved, it's mpossible to offer a step-by-step recipe for keeping bees with or without treatments. Most commercially available stock is abysmally bred and not selected for traits that the self-contained beekeeper would want. Beekeepers also face many environmental challenges that are out of their control, including factors that are guaranteed to impact bees, such as the presence of monocrop agriculture; the application of systemic insecticides; and spraying cranberries, soy, and other crops with fungicides while the bloom is open and bees are foraging.

Success and Failure Go Hand in Hand

A natural system, such as the one involving honeybees, flowers, and microbes, is dynamic. Genetic combinations are always being tested. Boundaries are always moving. A system with billions of interrelated components largely controls itself. Not every individual is fit. Not every population is suited to its environment. Not every combination or experiment is successful. This is the basis of any robust natural system.

The competition between components, including individual microbial strains, diverse bee genetics within the population, and the genetics and behaviors of the plants that provide forage, is fierce. It is also necessary in order to ensure that everything remains in balance. If one component fails or is too successful (i.e., a pathogenic microbe taking over and causing disease), the system breaks down.

Bees Are Not People

Bees are not like individual people and pets. Rather than concentrate on the health of any individual bee, queen, or even an entire hive, we need to focus on the health of the overall bee population. By letting nature cull the unsuitable individuals and colonies, we strengthen the population as a whole and avoid propping up any individuals or colonies that are unsuitable to breed. It really is this simple.

To Treat or Not to Treat?

The use of treatments (chemical or otherwise) represents a steep, slippery slope. Treated bees cannot be selected for bees that do not require treatments. Once your comb is contaminated with pesticides, it's a big, expensive, time-consuming job for you and your bees to change to uncontaminated comb.

If you're inclined to avoid using treatments, we urge you to start out without treatments and on small cell comb.

We can't guarantee that your bees will live through the winter if you don't use treatments. The truth is, many first-year beekeepers lose their bees over the winter regardless of whether or not they treat. The treatment-free beekeeping world is full of former treaters who lost their bees despite using everything they were supposed to.

If you want to avoid using treatments, then don't ever start using them in the first place.

The Least You Need to Know

- Diseases and pests are part of the natural system of keeping things in balance.
- Modern beekeeping practices encourage beekeepers to use several different kinds of chemical and antibiotic treatments on their hives.
- It's necessary for the long-term health of the bees to avoid using treatments on all diseases except AFB.
- The overall gene pool is much more important than the success or failure of a single hive.

Chapter 11

The Mating Scene

In This Chapter

- Where bees come from
- Honeybee genetics
- Getting good queens
- The whys and hows of requeening

Honeybees are unique in that the same bee can thrive in both feral (unmanaged wild) populations and as productive domestic producers of honey. The only difference is a little bit of management that redirects the bee's energy away from swarming and toward a honey surplus.

This ability of the bees to cater to both their own needs and the needs of humans is something that should be preserved in any breeding program. Unfortunately, much of our feral stock in North America has been decimated by habitat loss, exotic pests, and a failure of the beekeeping industry to recognize and make this need a priority.

Where Bees Come From

Almost all beginning, and many more experienced, beekeepers buy package bees in the spring from commercial bee suppliers. Where do the bees and queens that arrive in these 3-pound boxes come from?

Bees didn't always come in packages. Until about 100 years ago, bees, their heavy hive bodies, and delicate combs traveled together. Around 1880, A.I. Root (the commercial beekeeper, beekeeping author, and publisher) experimented with shipping the hive parts and bees separately. Packaging the bees meant that the heavy hive parts and foundation could be shipped unassembled at inexpensive freight rates while the lighter bees could be shipped express after the hives were assembled. Keeping the bees fed en route was the biggest obstacle, and the first reliable commercial shipments to the north occurred in 1912 (not by Root).

Bee Bonus

While A.I. Root was working to reduce the expense of shipping bees, he had no idea of the positive impact that the package bee industry would have on the prevention of transmission of American Foulbrood. Shipping bees separately from their combs meant that combs infected with foulbrood spores were left behind, and the bees would be installed on uninfected combs or foundation in their new location. To this day, some beekeepers successfully cure AFB infection by removing all comb and reinstalling the bees onto foundation.

Package bees give northern beekeepers a head start on making up winter losses. Package bees also contribute to an industrialized model of beekeeping, one in which worker bees are mass produced in the long seasons in warm climates, paired with unrelated (and unsuitable) queens, and shipped quickly and cheaply to distant locations—even between continents—in great numbers.

Bee Sex 101

We've already touched upon many of the basics of bee reproduction throughout the book, but now it's time to take a closer look at bee sex.

A few days after emerging and before mating, the queen will go on several progressively longer orientation flights. These flights introduce the queen to the environment around her hive and ensure that she will be able to find her way back when she finally mates. Because she is the longest bee in the hive and the slowest flier, she is in increased danger of being eaten by a bird or dragonfly. Drones are often observed escorting the queen on mating flights and are larger than the queen, so they make good decoys for predatory fliers.

Bee Smart

In order for a queen to mate properly, she needs to do so within a few weeks of emerging. Therefore, drones (emerged, and mature enough to mate) must be flying in the area. The best way to determine this is by seeing drones in your own colonies. These drones are an indicator that other drones in your vicinity are probably available for mating. There is no point to raising queens before there are drones available.

Drones seem to use the landscape to navigate to common aerial locations called *drone congregation areas (DCAs)*. Somehow, drones and queens from miles around all know where these DCAs are, and this is generally where mating takes place.

Queens sometimes mate in or near their yards, and an obvious concern is that the queen will mate with one of her own drones. A few things make this unlikely. For one thing, drones from multiple colonies are usually around a DCA, many from several miles away.

Bee Bonus

When a drone mates with a queen his sexual organ explodes, breaking off from his body, and he dies. Each subsequent drone removes the previously detached organ to gain access to the queen.

Drones drift freely from hive to hive and seem to congregate in colonies where there is a virgin queen or capped queen cell. Most of the drones in a hive are drifting visitors, not raised there, so even if the mating happens close to the hive, the risk of inbreeding is minimal.

Seeing some drones in the hive tells you that it's possible for queens to mate as long as it's also warm enough to fly. Seeing more drones in the hive tells you that the bees are thriving and can afford them. Many drones in the hive probably mean that there is a virgin queen. We should note here that finding all (or mostly) *drone brood* in a hive is a different story, and probably indicates a poorly mated or inbred queen. See Chapter 8 for more details.

A mating queen will take her flights until she has mated with enough drones to fill her sperm reservoir. Her body will keep the sperm alive, and she will use this sperm only to fertilize the worker eggs through her entire reproductive life, which can be up to five years.

Once the queen is fully mated, workers will feed her and fatten her up to the point where flying is difficult (and she has the nourishment to lay eggs). The queen will not fly again unless she is part of a swarm. If the hive is preparing to swarm, the queen will slim down by reducing her food consumption. It isn't clear if she stops eating or asking for food, or if the workers stop feeding her.

When a queen slims down before a swarm, her egg-laying activities slow or cease altogether. There can be a gap of several days between when the queen stops laying and the swarm leaves. After the swarm leaves, a virgin queen must emerge (in most cases she is close to emerging when the old queen departs). It will take several days for her to harden, take orientation flights, take mating flights, and start to lay. Altogether, this can mean a period of three weeks or longer in which no new brood is laid.

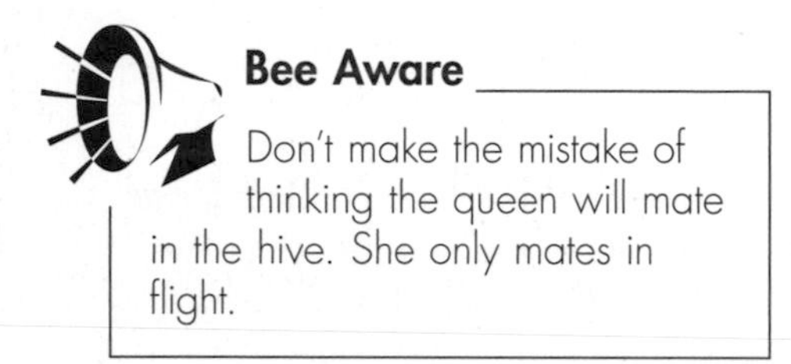

It's in the Genes

Mating habits and reproductive schemes of plants and animals tend to reflect their reproductive needs. Lobsters have millions of offspring with few expected to survive, while elephants rarely have more than one calf at a time. A lobster that had only one microscopic offspring at a time would go extinct very quickly, and if elephants bred like lobsters, there would be a lot to clean up!

When an animal has an unusual reproductive method, it's best to pay attention, as there are usually reasons why the system works, and there are consequences to "improving" things.

In order to understand the importance of genetics in breeding, we must take a closer look at the role of drones and queens in the mating process:

- **Drones** are produced in abundance when the colony can afford to raise them. Drone production usually indicates a strong colony, as they use up a lot of resources.

  Drones drift freely between colonies over an area of several miles. Since the queen of a colony will mate with up to 30 or more drones, odds are against a queen mating with more than 1 or 2 drones from any one colony. Each drone the queen mates with will only have genetic influence on a small percentage of the queen's offspring. This is a *wide but shallow dispersion of genetics.* It means that many colonies are influenced by the drones' genetic material (from any given colony), none to a huge extent.

- **Queens** are produced in very small numbers. At most, a hive will experience a few swarms and one or two new supersedure queens within a really active season. All successful queens will head up colonies, and all the offspring from each colony will have half the genetic material of their queen. Through rearing queens, bees in a natural system produce a few queens with a very *narrow but deep dispersion of genetics.* Few colonies are influenced by the queen's genetic material, but there is a strong influence in each of these colonies.

You must consider genetic influence when thinking about what kind of queens to use in your own apiary. A common beekeeping practice is to graft new queens from your best queen(s). In this way, you end up with a number of queen "daughters" from your best stock.

But grafted queens change the natural genetic dispersion. You now have the genetic influence of one queen spreading both wide (to many colonies) and deep (to all bees in all those colonies) from the queen side alone. Without a good deal of attention to the genetic history of the drones that the sister queens are mating with, such a scheme will very quickly lead to inbreeding. Care should also be exercised when purchasing mated queens. Deal with a queen breeder who is small enough to talk to you personally and able to assure you of the diversity of their stock.

Grafting can be useful when used judiciously and carefully, but *simple line breeding* is preferable, as it much more closely resembles what the bees do in nature.

Remember, every time you requeen with a mated queen from somewhere else, you are replacing the genetics of the hive completely. Always consider this when requeening. Do you really want to replace the genetics of that hive, or do you want to build on them?

def•i•ni•tion

Simple line breeding is when splits are made with brood or queen cells from the parent hive. It differs from a pure grafting method in that the progeny of many queens are used to head many splits, rather than using a few queens to head the same number of splits.

The Importance of the Queen

By the time you've read this far, you should appreciate the importance of the queen. Not only is she the mother of all the bees in the hive, but she does all her mating within the first two weeks of her life. Once a queen is mated and laying, the genetics of her offspring (and any traits/characteristics that are dictated by genetics) are fixed. Nothing can change the genetic makeup of the hive without replacing the queen.

Queen Production

Although a queen is as unique as a snowflake, she is the product of breeding, and the circumstances and conditions under which she is raised have great bearing on her qualities and performance. Her genetics, developmental nutrition, and mating all contribute to the health, vitality, and overall robustness of the generations of offspring she will produce. Each queen source has its own benefits, drawbacks, and cautions.

Raising Your Own Queens

Queens raised by you, or another small beekeeper you know, can be excellent. These queens might be grafted, produced with a "graftless" system (which forces the queen to lay in specially designed plastic cells for easy handling), raised from queen cells found by the beekeeper, or by any number of methods whereby the queen is removed and "emergency queen cells" are raised by the bees.

One caution is that queens raised in an emergency queen or grafting situation must be raised with abundant resources. This means plenty of pollen, plenty of honey, and abundant nurse bees. If you (or the person you are getting the queen from) aren't proactive in these regards, queens may very well be substandard. Nutrition is just as important as genetics for a great queen.

Emergency queens will be open-mated, and the genetic makeup of their offspring will likely be influenced by other bees in the area, be they ferals living in the wild, large commercial apiaries, or anything in between.

Queens may of course be raised as part of making splits or from maturing queen cells you find in the hive (for swarming or superseding). Just remember that if the bees are making queen cells, they probably have a reason. Take some measures to relieve the urge to swarm, or let them supersede.

Queens from Small Breeders

The small breeder is someone likely to have a reputation in your area for producing queens, and may or may not ship them. It's rare that someone with a limited production and good queens will have to advertise to sell out their supply.

Since a breeder with a good reputation isn't likely to neglect any of the obvious nutritional and nursing needs of raising queens, the questions you must ask revolve around genetics.

The first question to ask is where the breeding stock comes from. Ideally you will find someone who breeds from their own stock that they maintain themselves, only judiciously bringing in small amounts of new genetics at any one time.

Some smaller breeders simply purchase "breeder queens" from other breeders. Usually these breeder queens are artificially inseminated (which controls the variables that otherwise would be determined by uncontrolled outmating), and they are expensive (hundreds of dollars each).

When the breeder grafts from these "breeder queens," he or she is producing queens with a predetermined genetic makeup. The queen will be mated with whatever drones are in the vicinity of the breeder, and hopefully the breeder has made some effort to supply an overabundance of desirable drones.

The value of such queens really depends on the source of the breeder queen mother and the source of the drones the daughter queens mate with. A good breeder will have satisfying answers to questions about their stock regarding origin, breeding, and so on.

Queens from Commercial Breeders

The resources required to produce hundreds of queens a week are staggering. Each queen requires her own small hive (mating nuc), crowded with worker bees, while she is maturing and mating. The best breeders will leave the queen in the mating nuc for long enough to prove her ability to lay and that the resulting workers are healthy. That means that a mating nuc, once established, can only produce one mated queen a month.

Some suppliers take shortcuts and sell their queens before the brood emerges. You can be sure that suppliers that provide packages and queens to the majority of beginning beekeepers are not "proving" each queen before she ships out.

We recommend that if you start with package bees, you find a local, quality queen breeder to requeen with before the end of your first season. Queens provided by large package producers are bound to be substandard.

Bee Aware

In a survey done by the Barnstable County Beekeepers Association on Cape Cod, about 10 percent of packages had "drone layers" (queens that are not properly mated or inbred can lay drones and not workers), and 25 percent produced spotty (incomplete) brood patterns. Just under 50 percent had full brood patterns in six weeks. This is the nature of package bees, and not the fault of the beekeepers.

Requeening

Believe it or not, you are about to become emotionally attached to thousands of bugs. Your queen especially will reward you with prolific egg laying. Sadly, sometimes colonies need a new queen, and the old one must be dispatched in one way or another. As the beekeeper, you have the ability to choose when and how the queen is replaced.

Regime Change

Since the queen provides both the genetics and the egg production for the entire colony, her health and mating are of primary importance. Anything that interferes with the queens' ability to lay prolific numbers of fit workers is grounds for requeening. Reasons for requeening include the following:

- A hive has become overly defensive and is a danger to the public or the beekeeper.
- A hive is showing persistent symptoms of stress diseases (European foulbrood, chalkbrood, sacbrood), viruses, or mites.
- The beekeeper thinks that better genetics are available to replace current stock.
- The beekeeper wants to add genetics to his own stock.
- The queen isn't laying prolifically during a flow despite plenty of room, food, and young nurse bees.
- The queen has a general lackluster (or poor) performance.
- A queen is added to a queenless split (saving the hive a month of queen rearing).

Dethroning the Queen

If you've been paying attention, you've noticed that we've downplayed the idea of opening up the hive to find the queen, as she is the lady in the striped dress in a crowd of tens of thousands of ladies in similar striped dresses. The queen is the proverbial needle in a haystack.

Requeening is the only time a beekeeper needs to actually find (and remove) the queen from the hive. If you have trouble finding the queen, another set of eyes (especially those of an experienced beekeeper) can come in handy.

Finding the queen involves practice and luck (she sometimes does seem to just disappear). Starting with the most populated brood frames, examine each side of every comb. Look for a circle of bees surrounding her, and remember that if she is laying an egg you won't see her long abdomen. Start by looking in the center of the frame and spiral your eyes outward. Be patient and persistent, and stay as relaxed as possible. At some point she will just "pop out" from the mass of bees before you.

Bee Smart

An observation hive is an invaluable tool for practicing queen spotting. You can inspect the frames for as long and as often as you want to without disturbing the bees or worrying about being bothered by them.

If you cannot find the queen, or if the hive you are requeening is too aggressive/defensive to handle comfortably, you can break the hive up into single boxes, each with their own functional (if not permanent) tops and bottoms (make sure there is an entrance). The box that has eggs after three days is the one with the queen. You can divide them further if need be, and at some point, the population will be so low that the bees will be easy to handle and, with some persistence, you should be able to locate the queen.

Once you find the queen, you have to decide what to do with her. You can use her to start a small nucleus colony (see Chapter 4), or you can pinch her ("pinch" is beekeeper speak for "put to sleep"). If you really don't have the nerve to use your fingers, you can use a hive tool to dispatch her. The queen should be removed from the hive for at least a few hours (preferably 24 and not more than 48) before requeening.

Bee Smart

Remember to save the pinched queen in a small bottle of alcohol for a lure tincture.

Requeening Methods

The bees know almost immediately that the queen is gone, and if there is young brood present, they will start preparations to raise their own queen. It will take a month before a new queen can be raised and mated and another three weeks beyond that before new brood can emerge. In the meantime, the colony is losing an entire month's worth of brood and the labor they can provide. Requeening offers the alternative of introducing a laying queen who starts working where the old queen left off.

Nucleus colonies can be used to requeen and boost a hive at the same time. Here's what you do:

1. Remove the cover of the queenless hive, and place a sheet of newspaper over the colony.
2. Place a hive body (with either foundationless frames or frames of small cell foundation) above the newspaper, and rip a couple of small holes in the paper with the hive tool.
3. Install a nucleus colony in the (now top) box just as described in Chapter 7.
4. Replace the cover and leave the bees alone.

The newspaper slows the mixing of the bees and, as they chew through the paper, they should become one happy family.

If you've purchased a queen, she probably is in a cage suitable for introduction. Follow these steps:

1. Either hang or squeeze the cage between combs, making sure that the bees in the colony can feed and groom the queen through the mesh or screen of the cage.
2. After three or four days, remove the cage. If the worker bees are clinging and biting the cage so hard that it is difficult to brush them off, replace the cage for another few days.
3. Release the queen directly onto the top bars (so she can climb down between the combs) only after the bees are not aggressive toward the cage.

Bee Aware

Remember that there can be two laying queens in a hive. If the bees are reluctant to accept a new queen, look again for eggs. There might still be a queen in the hive.

The most dependable way to introduce a queen to a queenless hive is with a push in cage. The cage can be purchased, or made from #8 steel mesh. Here's what you do:

1. Locate a frame with capped and emerging brood in the queenless hive. If there are no such frames in the hive, then take one from another colony (you can shake the bees off before transferring it).
2. Brush the bees off the frame and put the queen in the push in the cage directly over emerging brood. Secure the cage making sure there are no other adult bees in the cage. Don't worry if you damage a few brood in the process.
3. Replace the frame in the hive and close things up. The newly emerged bees (now in the cage with the queen) will accept the queen and help spread her pheromones to the rest of the colony, and the queen can lay in the vacated cells.
4. In four days, check to see if the bees are still clinging to the cage. If not, release the queen. If so, give her another couple of days. Look for signs of another queen in the hive if they are reluctant to accept the new queen.

Virgin Queens and Queen Cells

If you want to have your new queen mate locally, or if you really cannot find and remove the queen, you can try using a queen cell or a virgin queen to requeen your hive.

A queen cell (sourced from a breeder, or a capped queen cell you find in another hive) can be placed outside the broodnest either on the frame the bees built it on, or in a specially designed cage. The virgin queen will likely emerge and supersede the existing queen.

A virgin queen (preferably emerged in an incubator) can usually enter a colony unnoticed and supersede the old queen. Give the bees a very heavy smoking (so that smoke blown in the bottom comes out the top), and then release the virgin queen right into the entrance.

Don't be hesitant to requeen a colony when you are convinced that there is better stock available, but don't get caught in the trap of constantly trying different genetics. It will take some time for your stock to sort itself out, and you have to give your efforts a chance to work before "fixing" them.

The Least You Need to Know

- The genetics of the queen and drones she mated with absolutely determine the genetic makeup of the hive.
- Package bees do not come with the best queens.
- Requeening with a mated queen completely changes the genetics of the hive.
- If you purchase queens, deal with a reputable queen breeder, preferably a local one.

Part 4

Harvesting and Beyond

Honey is more than just a treat for the beekeeper. It is the bees' pantry, intended to help them survive the winter. You cannot harvest willy-nilly, and must take only surplus honey that the bees can survive without. You'll learn various techniques for harvesting honey and what to do with it once you have it.

Once you've harvested your honey, it's time to start thinking about your winter plans. Winter is a great time to prepare for the following spring, put together new equipment, read, and yap about bees on the Internet when everyone else in your life is sick of hearing about them.

No doubt you will be daydreaming about cutting your hours at work, finding more prime locations for your hives, and scheming how to obtain more bees (and keep them hidden from your partner). There are all kinds of ways to generate income from keeping bees—producing and selling honey is only the most obvious of them. Teaching others about beekeeping, be they elementary school students, apiary visitors, or aspiring beekeepers, is extremely rewarding.

The life of a beekeeper offers many pleasures, not the least of which is the privilege of spending time with the bees themselves as their captivating behaviors unfold, and they perform the alchemy of turning straw into gold.

Chapter 12

Honey, Harvest, and Resources

In This Chapter

- The beeconomics of honey
- Methods for harvesting honey
- Beeswax as a product
- Making your own foundation

We know we've been going on and on about the bugs, and all you want is the honey. It's finally time to talk about one of the sweetest rewards of keeping bees.

Honey Is Money ... and Comb Costs

Not only does it take honey to produce comb, but until the comb is built, no honey can be produced. By the time the comb is ready, the flow might be over. For the bees and the beekeeper, comb is an investment that, in most cases, must be made well in advance of a crop.

Beeconomics of the Colony/Beekeeper Relationship

Surplus honey is a reward, one that consumers have come to expect as a matter of course, like strawberries in December. Toward this goal, it isn't uncommon for beekeepers to feed 100 pounds of sugar to their hives in order to turn around and harvest 100 pounds of honey from them. It seems to us that such an artificial method of obtaining honey is not what keeping bees is all about.

Purpose for the Surplus

Bees, when left to their own devices, use a surplus of honey to swarm and reproduce on a colony level. As a beekeeper, your goal is to redirect this behavior so the bees produce enough surplus honey for you to harvest. In exchange, you allow the bees to reproduce at a controlled level and you provide secure housing for the parent colony and its splits so that it doesn't need as many offspring.

Where Honey Ultimately Comes From

Producing honey is not something you can achieve with brute force. Instead, honey requires gentle intervention. It all comes down to directing sunbeams, in the form of solar energy, into the jar.

In rough terms, plants containing chlorophyll use sunlight to convert carbon dioxide and water into diluted plant sugars. Plant sugars are essentially energy that can be stored as sap, as fruit, or expressed as nectar. Honey is, quite literally, stored sunshine.

Plants use sugar in the form of fruit to attract animals, which will spread the seeds after eating the sweet fruit. The other major use of sugar by plants is to entice pollinators to visit their nectar-filled flowers. In the process of gathering the nectar, the insects also pollinate plants, thereby aiding in reproduction.

Bee Bonus

Another type of plant sugar is maple sap. Maple sap is much more dilute than nectar but the big difference between maple syrup and honey is that with maple syrup we can appreciate how much energy is required to concentrate it enough so that it won't spoil. Bees do almost this exact job with nectar, except they use their wings instead of fire.

Harvest: Separating the Sweet from the Sting

Bees know their tasks, and they do them well. Bees rarely cap honey before it has reached about 18 percent moisture, which is what it has to be so that it doesn't ferment. Honey is ready to harvest when the entire frame is capped. If a frame is 90 percent capped, it is probably okay to extract the honey, as even if the uncapped honey is a little wet, the capped honey is likely to be drier, and when things mix in the extractor, the batch should even out to 18 percent or less. If you're in doubt over the moisture content of the honey (and bees do sometimes cap unripe honey), you can use a tool called a refractometer to measure the moisture content.

Once you have identified frames to harvest, you have to wrestle them away from the bees.

Bee Aware

We have heard several stories from friends who brought honey back from some exotic vacation spot in a coconut shell, and weeks later, the coconut exploded from the gas built up from fermentation.

Commercial Systems

In most commercial operations, queen excluders are used to keep brood out of honey supers. Bees can be removed from an entire box with either a bee blower (which, as the name implies, uses a big fan to blow the bees out of the box) or with a chemical repellent (repellent is applied to a fume board and placed above the boxes that the beekeeper wants to harvest. Either method drives bees out of the upper boxes and into the broodnest, leaving the box empty of everything but the honeycomb.

Bee Smart

How much honey you can take from the bees depends greatly on location and time of year. We don't recommend harvesting honey until a hive is self-sufficient and has surplus above the third deep box (or its equivalent in smaller boxes).

Honey Supers

There are two less intrusive ways to clear bees from a box of honeycomb. One involves a device called a bee escape, which is a one-way door that is placed below the boxes that are to be harvested and left in place overnight. In theory, as the bees leave the box, they can't get back in. This works best when evening temperatures are a bit lower, as the bees tend to leave the honey supers to cluster around the brood at night.

We offer one important word of caution. If there are any places where bees can get into the super from the outside, you will likely find the super robbed out, often by the very hive upon which it sits. When you lift the box, it will seem too light, and the honey will be gone.

The second way to remove bees from a box of honey is to wait until dusk. Just as it starts getting dark outside, remove the super and replace the cover on the hive. Stand the super up on its end (with frames perpendicular to the ground) behind the hive. As darkness settles, most of the bees will fly back to the hive, and you have a virtually bee-free box to deal with. Make sure to remove the box to a bee-proof location as soon as the bees leave to avoid robbing.

Unlimited Broodnest

All along we've been advocating that you adopt an unlimited broodnest approach to managing bees. Harvesting honey with unlimited broodnest presents its own set of issues, the most notable being that you are not assured that any box will be free of brood. For this reason, you'll need to harvest honey frame by frame. Choose frames that have only capped honey. Leave the frames with honey and brood for the bees.

Remove frames one by one from the hive, checking for capped honey. One at a time, using a bee brush or a wad of hay or grass, brush the bees off any frames that are to be transported back to the kitchen or honey house for extraction, and place the frames in an empty hive box. Keep the box covered except when adding a frame to the box.

With this system you will have the extra work of sorting frames (as opposed to being able to take a whole box) but frame-by-frame harvesting means that lifting is greatly reduced.

If there is not a flow on, the bees will be very interested in the honey you are harvesting. To avoid having the bees rob you of honey, keep the lid bee-tight. For particularly pesky bees, try covering the box with a damp sheet; it should be heavy and clingy enough to keep the bees out.

Bee Smart

Place harvested honey frames in five-frame nuc boxes to reduce the weight of each box you have to carry.

There will doubtless be a few bees on the honeycombs in the box. Simply let them out after an hour or so, and they will be happy to fly free.

After you transport the frames full of honey to your kitchen or honey house, it's time to cash in on the liquid gold. How you do this, once again, depends on your goals and circumstance.

Cut Comb

The easiest way to enjoy honey is by making cut comb, which, as the name implies, is simply honeycomb cut into pieces. There are a few details that need to be attended to for success:

1. Lay the frame on a bakery cooling rack or a screen placed over a tray with raised sides. Using a knife, cut the wax out of the frame. Don't cut too close to the top bar—leave a row or two of cells for the bees to redraw the comb.
2. Cut the comb into whatever size you desire, which more than likely depends on what size containers you have to put the pieces in. Honey will drip from the cut edges of each section. Let them drain overnight so that the honey does not settle around the comb, which looks less attractive.

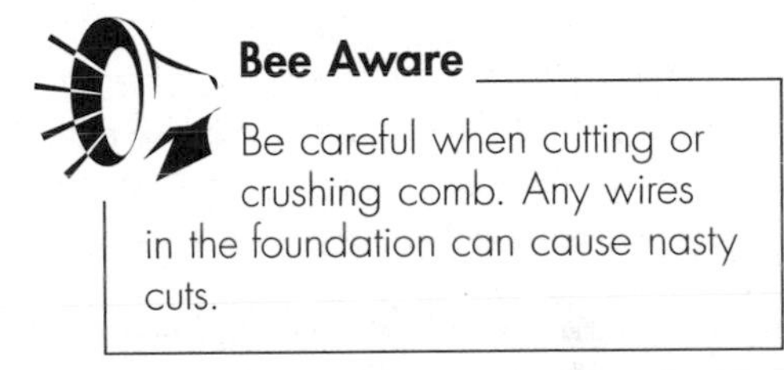

Bee Aware

Be careful when cutting or crushing comb. Any wires in the foundation can cause nasty cuts.

3. Place the drained pieces of comb into the desired containers.
4. There is always a possibility of wax moth eggs being present in the comb, and they will destroy your packaged product if they hatch. Therefore, once packaged, freeze each cut comb section for three days. This will kill any eggs, and it will not harm the honey. After three days, the cut comb can be thawed and eaten.

Eating comb honey is a real treat. The honey has not been aerated via extracting or crushing, and it is as close to sampling honey in the bee yard as you can get without the risk of being stung.

Enjoy the comb straight with a spoon, or use a fork to uncap the top side of the comb, and use a spoon to scoop the honey out. The comb can be flipped over to enjoy the other side.

Bee Smart

There is no nutritive value in beeswax. It is perfectly safe to eat clean wax from untreated colonies, but your body will not process it. If your goal is to produce cut comb honey, special extra thin foundation is available for this purpose. However, given the current data on beeswax contamination, we do not recommend using foundation for cut comb. In a recent test performed by researchers at Penn State, five sheets of foundation were sourced from five suppliers. All five had high levels of fluvalinate and coumaphos, things that you do not want to ingest (see Chapter 10). If you want to produce cut comb honey, we strongly recommend foundationless frames.

Crush and Strain

The crush and strain method allows you to produce liquid honey without having to use any expensive equipment. Here's what you do:

1. Buy two plastic pails. In one pail, drill several ½-inch holes in the bottom, and line the pail with a mesh nylon bag. (These are available from any beekeeping supply house.)
2. Place the pail with the holes and liner inside the other pail.
3. Remove the comb from the frame the same way as when producing cut comb.
4. Place the comb in the mesh bag and crush the comb so that each and every wax cell is ruptured. The honey will run through the mesh bag to filter out the wax and drip into the bucket below.

Depending on the honey, it might take overnight or longer for all of the honey to drip into the bucket, and this technique will work much better at a temperature of at least 80°F.

Bee Aware

The naturally acidic nature of honey reacts with aluminum, etching into the metal and ultimately resulting in contaminated honey. The best materials to use when dealing with honey are stainless steel and glass, both of which are nonreactive.

If you are going to do this regularly, inexpensive setups are available from bee supply houses, or you can install a plastic "honey gate" on a five-gallon bucket for easier transfer to a bottle or other container. Some beekeepers use cider presses or automobile jacks to put great pressure on the crushed comb, forcing the honey out.

Smaller quantities of honey can be crushed and strained in a smaller vessel or colander, but make sure not to use anything aluminum with honey.

After you've gotten as much honey as you can out of the wax, give the sticky wax to your bees for easy cleanup. If it's not convenient to give it to them immediately, store in an airtight container until the time is right. Feed either in the hive (under the cover), or at least 50 feet away from the hives; otherwise, you might induce robbing.

You can save any clean, untreated wax for your own use or to sell or trade to anyone who is looking for clean wax for foundation making, cosmetics, or salves.

Extraction

An extractor is basically a device that spins the honey out of the cells of the combs using centrifugal force (think "spin cycle" on a washing machine). There are two major types of extractors, radial and tangential. The difference between them lies in the orientation of the frames to the axis of the extractor. In a radial extractor, the frames radiate out from the center, like spokes in a wheel. In a tangential extractor, the frames are oriented perpendicular to the imaginary spokes.

Bee Bonus

The honey extractor came into use in 1865, and paved the way for modern honey production. For the first time, honey could be harvested and the precious comb returned to the hive to be refilled. Essentially, the extractor spins honey out of the comb.

Radial extractors can extract honey from both sides of the comb at once. However, radial extractors are not as good at getting very thick honey out of the comb (as the cells are near perpendicular to the centrifugal force), and they also require higher rotational speeds. Most of the current high-end extractors are radial, and they are considered superior for most applications.

Tangential extractors do a better job of getting the honey out of the cells (as they are parallel with the centrifugal force), but the frames have to be flipped. To do the job right you must remove about half of the honey from one side of the comb, flip the frame and extract the honey from the other side, then flip the frame back around and finish extracting the honey from the first side.

For foundationless frames, we strongly recommend tangential extractors, as the comb is supported by the frame basket (a wire enclosure that holds the frame in place but allows the extracted honey to pass).

Bee Smart

Extractors, even used ones, are not cheap. Do not feel compelled to purchase one with your initial order of bees and equipment. Most bee clubs have an extractor that members can borrow or rent. You are better off getting by without an extractor until you can justify buying a high-quality one with a motor that can handle more than four frames at a time. There is no comparison between top-notch professional extractors made in the United States and cheap imports.

Before extracting can begin, you need to remove comb cappings using a capping scratcher, an uncapping knife, or an uncapping plane. The capping scratcher, sometimes called an uncapping fork, is a simple tool that can be used to either scrape the cappings enough to allow the honey to be extracted or to pry the cappings off the cells.

Uncapping knives and planes contain electric heating elements that melt through the comb. The comb is usually a bit wider than the top bars, and the knife or plane trims it back to be level with the top bar.

Capping wax is usually light in color and clean. After the honey is removed from the cappings (by spinning them out, by heating, or by feeding to the bees), it should be saved as premium, treatment-free wax.

Into the Jars

Extracted honey can be stored indefinitely in any clean glass or stainless-steel container with a tight-fitting lid. Cleanliness is important when handling any food, but honey has natural antimicrobial properties, so no sterilizing is necessary. Keep honey covered at all times, as it can draw moisture from the air and begin to ferment. Honey is best stored in a cool place—preferably above 60°F and below 90°F—and out of direct sunlight.

All honey will crystallize over time. Once you experience raw, crystallized honey, you may enjoy it more than liquid honey—we certainly do.

However, if you need to re-liquefy honey, put the container in a pan of hot tap water so the water comes to just below the lid, and wait. Do not put the honey in the microwave or expose it to high heat—anything over 100°F is too hot. Microwaving can create hot spots and any excess heat will denature enzymes and destroy subtle flavors.

Raw (unheated, unfiltered) honey imparts different flavors at different temperatures. Even at temperatures below freezing, honey never completely solidifies. Enjoy at any temperature up to 100°F and take note of the varying taste sensations.

Beeswax as a Product

Beeswax has been recognized for its valuable qualities for thousands of years. It has been used in numerous products, including candles, cosmetics, furniture polish, sewing aids, metal casting, bullet lubrication, candy making, batik, and musical instruments.

If your primary goal is honey production, then extracting the honey from the combs and reusing the combs is the most efficient method. However, if you want to produce wax in any quantity, and are willing to sacrifice some honey production, crush and strain is a better approach.

Contaminated Supply

The beekeepers who supply the most wax are those with the most hives, which are also the ones who tend to treat the most. Think about it: if you have 50,000 hives, you don't evaluate each one to see which ones need treating; you treat them all. The contamination of the beeswax (with fluvalinate and coumaphos especially) seems to be ubiquitous. Essential oils are also readily absorbed by beeswax.

Bee Aware

It's general knowledge that large commercial beekeepers use all manner of unapproved treatments that are not even being tested for. One of the major cosmetic companies purchases all of their beeswax from third-world countries so as to minimize such contamination.

Clean Wax

If you want clean wax, you must either find a beekeeper who doesn't treat his or her bees, or you have to become that beekeeper. If you use foundation, you will introduce some contamination into the hive, but not so much that we recommend against using it at all. We do not, however, recommend producing cut comb with foundation. Instead, use foundationless frames and let your bees draw all the wax. Without the midrib of foundation, your comb will be more tender and delicious, resembling honeycomb produced in the wild.

Making Use of Wax

Today's beekeepers tend to use beeswax primarily for candles, foundation, and cosmetics. We don't recommend wasting uncontaminated beeswax on candles, as clean wax is too rare to simply burn it away. When making cosmetics and foundation, on the other hand, you really want wax that is as clean as possible.

There are all manner of beeswax hand cream, lip balm, and soap recipes available—but we won't include them here. Commercially available preparations are not in the same ballpark as what you can make with quality ingredients at home. Clean beeswax is essential if you are going to use homemade products on your body. Would you really want to use or sell a lip balm with even small amounts of organophosphates (see Chapter 10)?

Making Your Own Foundation

Making your own foundation is not as hard as you might think. Basically, you soak a board (slightly larger than the size of the foundation you want to make) in water to prevent wax from sticking to it and dip it in molten beeswax. When the wax hardens somewhat but is still pliable, you trim the four edges of the board and remove the sheets of wax from the two broad faces of the board. While the wax is still warm, you run it through an embossing mill, which is essentially a pair of rollers with a negative of a cell impression, and then trim the foundation to size.

Embossing mills are available in several cell sizes. If you are making small cell comb (which is what we recommend) then you want a 4.9mm mill.

Making your own foundation can be very expensive. The embossing mills alone can cost $2,000 or more. Plus, most methods require an enormous quantity of wax. It would take several years for a beginning beekeeper to accumulate enough to make even a few sheets of foundation, and even then, many pounds end up sitting in a reservoir that can't be used without adding more wax.

Bee Bonus

We are currently developing a system for making your own foundation that requires much less wax and includes all of the needed equipment. The goal is to provide an affordable way for beekeepers to produce their own foundation with minimal investment. This is a work in progress, and updates and information will be available on our website, www.beeuntoothers.com.

The Least You Need to Know

- Honey should not be harvested until a hive is stable and showing a true surplus.
- Honey can be harvested with cut comb, crush and strain, and extraction methods. Extracting honey from the comb enables you to reinsert the comb for bees to reuse.
- The cleanest wax available is from your own untreated bees.
- Making your own foundation is an important step toward a clean, self-contained operation.

Chapter 13

The Offseason

In This Chapter

- Why winter begins in summer
- Staying busy throughout the winter
- Emergency winter feedings
- What to do when spring comes around

Contrary to what you might expect, the offseason for beekeeping begins in the middle of summer. By July, you should be thinking about what your bees will need to survive the winter and start making any necessary preparations.

Size of the Colony

The size of the colony is an important consideration. It's true that small colonies can be overwintered even as far north as the Canadian border, but this is generally accomplished with specialized equipment, extensive preparations, experience, and specific goals in mind.

You are going to need the equivalent of at least two, but preferably three, deep 10-frame boxes for successful overwintering in areas that have a real winter.

If your bees don't draw comb during the summer months, it is unlikely they will do so come fall when their needs are more immediate. Without comb, the bees can't cluster properly, can't store food, can't regulate the airflow in the hive properly, and can't do all the other things they need to do in order to survive.

Layout

Just as important as the number of combs is their arrangement in the colony. As fall approaches, empty small cell combs should be maintained in the core of the broodnest so the queen has a place to lay, and honey stores and pollen should be accumulating outside the broodnest. Don't overmanipulate the frames, as the bees generally know what should go where. Your job is to keep the broodnest open so that the bees don't swarm late in the season. Once temperatures get colder—consistently under 50°F at night—you should cease moving frames unless you are putting honey frames next to the broodnest.

Bee Aware

Never move frames unless you have a solid reason for doing so. If you are in doubt, leave things the way they are. Overmanipulation of frames is the cardinal sin of the new beekeeper.

Stores

In New England, in addition to however much pollen the bees store, you should leave at least 100 lbs. of honey (150 lbs. is better) with the bees for winter. Surplus honey stores give them capital with which to get a running start on spring. With these resources, they won't have to wait for weather and forage to cooperate before they can start building up for the season. A head start is invaluable come spring, and will eliminate a lot of feeding and coddling that you would otherwise need to do.

Leaving the bees with stores of this quantity is unpopular among many beekeepers simply because they want to harvest as much honey as possible before winter. We feel this is short-sighted and contrary to the nature of the bee.

Ventilation, Not Insulation

Many people mistakenly assume that insulation is the most important factor for helping the bees get through winter. Not so. In the winter, ventilation is much more important than insulation.

Bees don't hibernate, but they do display interesting behavior in the winter. Their main activity during cold weather is clustering. Bees are able to disengage their flight muscles from their wings, and they generate heat by flexing those powerful muscles—basically turning themselves into honey-powered space heaters!

Even in the dead of winter, they are able to raise some brood (which requires a constant temperature of 94°F!) thanks to their ability to cluster. When bees on the outside of the cluster get cold, they move toward the middle, and when those in the middle get too hot, they move to the outside.

As long as strong wind isn't moving across the cluster and they have enough food, the bees are able to maintain this state all winter.

Bee Bonus

On any warm day, bees will take a cleansing flight. Look for yellow spots in the snow, on your drying laundry, or your car. Anything in the flight path of the bees is fair game. Not every bee will make it back from such flights, and you will generally find at least a few dead bees in the snow. This is normal and nothing to worry about.

While the bees are in a cluster, warm, moist air from the bee's respiration rises in the hive. Without proper ventilation, when this air hits the top, it will cool rapidly, and water will condense on the inside of the top cover. If it's cold enough, the moisture will freeze there above the cluster, only to drip down and freeze the bees when things warm up. If it doesn't freeze, it will drip down immediately, killing the bees.

A top entrance or vent is essential so that warm, moist air can escape the hive before it condenses. You also need a vent in the bottom of the hive to replace the air escaping through the top vent.

The only insulation that's necessary is over the top cover to further minimize condensation. A standard telescoping cover with an inner cover will have an insulating effect. Otherwise, a sheet of rigid foam or a plastic bag full of dry leaves will work. Don't forget to secure the uppermost layer with a brick or a good-size rock.

Winter Inspections

You'll be limited in how much you can observe the bees in winter. On warm days you should see some bees taking cleansing flights. This is one of those times when having multiple hives is helpful; if three hives are flying and one isn't, you know to look for a problem. You can gently lift the back of the hive to feel for weight. If it feels light, you can consider a quick inspection and emergency winter feeding. Sometimes colonies that don't have sufficient stores will fly in desperation.

Working as quickly as you can, remove the top and any boxes over the broodnest. Inspect the broodnest. If the bees have no food stores, you need to feed them in a way that they can access the feed without breaking the cluster. You can try punching a few holes in a paper plate, filling it with crystallized honey, and placing it over the broodnest. If clean honey isn't available, lay a sheet of newspaper covered in dry cane sugar over the broodnest, drizzle a little water at the edge of the sugar, and let the mixture drip down into the hive (a few drops, not a gallon) so that the bees sense that the sugar is there.

The bees can chew up through the plate or newspaper, and access the feed as necessary. Keep in mind that this is for emergency use only! If you find that you are always doing such winter feeding, you are not starting the winter off with nearly enough stores for the bees.

You should never do a full hive inspection in the winter. Going through the hive frame by frame when it's cold outside will probably kill the bees unless you have experience doing such inspections.

Making the Most of Winter

If you live in a climate that has any kind of winter, you may go from four to six months without spending time with your bees. This stretch can feel endless. Fortunately, there are many winter activities that can keep you connected.

Looking Back and Planning Ahead

Winter is a perfect time for thinking back on the past year and making plans for the next. What went really well? Are there areas where you may want to make changes next year? Will you need more equipment? More information? Would you like to conduct any experiments? Take some time to reflect, set some goals, and make plans for how you'll spend these next few months.

Building Equipment

Even if you are happy with the number of hives you have, it's always a good idea to have some backup equipment when spring comes. You may want to capture some swarms—either yours or someone else's—or split your hives. You may also want to purchase more bees. It's a lot easier to have equipment ready when the need arises than to patch things together at the last minute or after the swarm has disappeared.

If you have access to cheap lumber, it may be worth your while to construct your own hive bodies. We have found that it is cheaper for us to order pre-cut woodenware for boxes and frames than to purchase the raw wood and mill it ourselves.

Don't forget to include a few extra tops, bottoms, and hive bodies in your building plans. You never know when you'll need them.

Bee Smart

If you can get a group of beekeepers together and order woodenware by the pallet, your shipping costs will be greatly reduced. If you have the capital, you can order in bulk and sell the surplus to beekeepers in your area. Remember, with 3 to 4 deep boxes (or their equivalent) per colony, it will only take about 12 to 16 colonies to use up an order of 50 boxes.

Ordering Supplies

Beat the spring rush and order next season's bee supplies in the winter. You'll have all your equipment built out and ready to go while everyone else is on hold with back-orders. Deciding what you need ahead of time also means you'll know what to put on your holiday wish list.

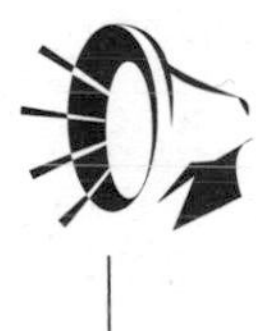

Bee Aware

Suppliers don't ship foundation in the winter, as cold temperatures make the wax brittle and vulnerable to breaking. If you want to embed foundation over the winter, be sure to order it in the fall.

Get Smarter

What better way to spend a cold, snowy winter day than curled up with a hot mug and a great book? There are dozens of wonderful bee books begging for your attention. You could also curl up with a laptop, as the Internet is loaded with almost anything "bee" imaginable. Research journals from around the world, beekeeping forums, used book sites, blogs, informational websites, even mp3 recordings of seminal beekeeping texts are all available online.

Self-education is important for any beekeeper because there is so much to know and explore, and there are so many conflicting opinions. Every beekeeper has his or her own unique perspective and ideas about how to be a better beekeeper … you will, too. Besides, learning more about bees is just plain fun. Be the master of your education!

Bee Smart

Many journal articles are available online, but you may be required to pay the publisher to download each one. If your local public library is unable to help you, try a college or university library in your area. Same goes for books: many public libraries are part of inter-library systems and can find obscure books for you. We have been able to borrow $300 textbooks by simply having an active library card.

Building Community

If you haven't made a connection with any beekeeping communities yet, winter is the perfect time to do so. People miss their bees and are usually happy to get together with other beekeepers, either virtually or in person. Check out the online beekeeping forums and find your local beekeeping club. Many beekeepers belong to more than one, which increases their opportunity for networking and potlucks! Every state has an agricultural extension service where you can locate clubs in your area.

Internet forums abound, and you'll surely find one suited to your personality, level of knowledge, and intensity. Thanks to web-based language translation software, you can even communicate with beekeepers who don't share your native language. The Internet has brought the beekeeping community together despite geographical, cultural, and social boundaries. Being able to share information and ideas in real time, with your beekeeping neighbor or beekeepers around the world, provides unparalleled opportunities for learning.

Beekeeping conferences often take place during the winter and are great opportunities to meet your virtual friends in person and to make new friends. Conferences focused on treatment-free beekeeping are increasingly happening around the country and are wonderful sources for sharing knowledge, friendship, and enthusiasm.

Bee Aware

Conferences can be great ways to connect with other beekeepers but they can have their drawbacks. If you are coming from a treatment-free perspective, you may be shocked at how much of the conference content is focused on treating mites and disease. For those of us on natural systems, for whom mites and disease have ceased to be major issues, this focus is frustrating and depressing.

More Bees, Please

While we all hope that all of our hives will overwinter successfully, some losses are inevitable. You may be able to make up any losses by splitting in the spring, but you should also be prepared for the possibility that you'll need to replace bees from outside sources until you build up an operation large enough to be self-sufficient.

Spend some time networking online and in your local beekeeping community for possible suppliers of locally bred, untreated bees. Note that anyone who has bees you might want will sell out early. It may be worth it to reserve some bees in advance, whether or not you are sure you will need them. If you don't end up needing them and they are quality bees, the supplier probably has a waiting list, or you will easily find another beekeeper to resell them to.

Spring Comes

As spring comes upon your bees and the weather gets warmer, you will be anxious to get inside the hives. Be patient. You can get a bit more involved, but you should not pull brood frames or keep the hive open for very long until the temperature hits 70°F.

Early Intervention

With that said, if you think the bees are having trouble accessing food in the hive, you can move some empty comb away from the cluster and add full honeycombs abutting it. Be careful, as the cold comb, whether empty or full, is brittle, and if you mishandle it, will break and shatter.

If you notice extensive streaking on the outside of the hive, your bees likely have some form of dysentery. Dysentery can be caused by a number of factors, including being confined for too long, nosema, or other intestinal maladies. Usually, making sure the bees are in full sun, have good ventilation, and have plenty of food will clear this up.

Once dandelions come up in your area, the bees should start building up noticeably. There are earlier sources of forage, including maple, skunk cabbage, and willow, but the dandelions are the first abundant forage after the temperature gets high enough for the bees to really go to town.

You can now start to use some of the equalization techniques we discussed in Chapter 5 (moving frames of brood and/or honey, switching locations of strong and weak hives, etc). Just don't spread your resources too thin and handicap your best bees.

Splitting

You can start making splits once the bees have reached critical mass—usually more than six frames of brood—and you see emerged drones in your hive; if there are no drones, new queens can't mate. Remember that most of the time you are focusing either on making bees or making honey, not both. If you are intent on getting honey, do not split all of your strong hives as soon as they reach critical mass, as the strong hives will have the large populations needed to bring in honey.

If you lose hives over the winter or early spring, try to determine the cause. Most anything short of American foulbrood can be repopulated with new bees or splits from your other hives. If frames are full of dead, diseased brood, it's best not to use them. Adding nucs, splits, or packages to drawn comb is a lot easier than starting out new with no drawn comb. With drawn comb available, the bees will come on faster than you can believe. Remember the bee math. Under ideal conditions, the bees can double the amount of brood every three weeks. Keep this in mind as you make splits and build up for the season. If you have 6 frames of brood in a single box, make sure you add more boxes so that three weeks later you have room for 12 frames.

Spring is the most exciting time for a beekeeper, and probably for the bees as well. The whole season stretches out before you with the potential of honey, more bees, and more adventure than you can imagine. You are now into your second year with more experience and knowledge than you had your first, yet somehow there seems to be so much more to learn.

We are reminded of a quote from Karl von Frisch, who discovered the dance language of the honeybee:

"The bee's life is like a magic well: the more you draw from it, the more it fills with water."

The Least You Need to Know

- Comb is essential for successful overwintering and needs to be built early in the season.
- Beekeepers are almost as interesting as bees.
- Build equipment before you need it.
- Have a plan to replace your bees in case they do not survive the winter.

Chapter 14

Beyond Your Own Backyard: Building a Bigger Operation

In This Chapter

- Products of the hive for sale
- Where to market your products
- Placing bees on other people's property
- Living the beekeeping life

By now, if you are anything like us, you are obsessed. Unfortunately, the cure is the disease … more bees!

A Business of Bees

When you produce more honey than you can eat, it's time to start thinking of your hobby as a business.

Honey Sales

Selling honey is an obvious way to earn money from beekeeping. Quality honey sells itself, and for good prices. Farmers' markets, festivals, stores, flea markets, the Internet—pretty much anyplace there are people is a good place to sell honey.

Bee Smart

The most important part of selling honey is giving out tastes. Even though honey is antimicrobial and not of concern for contamination, it's important to keep things sanitary to the standards of the local health department and your customers. The old saying, "It's better to ask forgiveness than permission," holds true when it comes to offering samples. Do things in an exceedingly sanitary manner so that if any inspectors happen by your stand, they are reassured by everything they see.

With a premium product comes premium price. Treatment-free honey is an exceedingly rare thing on the market. Most Americans have never tasted real honey and will be surprised at how delicious and different it is from what they are used to. Tasters who were born in many other countries will get dreamy looks on their faces and begin to reminisce about their childhoods and the honey they ate in their homelands. If you are willing to educate your customers, give tastes, and simply talk to them about bees, you will sell honey at any price.

The secret weapon for the beekeeper selling honey or any other bee-related products is an observation hive. Bees behind glass with a laying queen and bees emerging from their cells will be the focal point of any market, particularly if there is a beekeeper around to educate people about what they are seeing.

Manufactured Products

Value-added products are another avenue of income. Lip balms, hand creams, candles, soaps, or foods containing honey are all easy for the beekeeper to sell. Although health department regulations are generally lax or nonexistent with regards to hobby beekeepers selling honey they have extracted and bottled in their own kitchen, the same is not true of any processed product, including honey purchased from other beekeepers. Before you sell such items, check with your health department and find out what you need to do to be approved and licensed. It's also worth looking into a product liability (and perhaps bee liability) insurance policy.

For the Bees

For some beekeepers, producing bees is more profitable than making honey. If you have bees with which you can demonstrate survival and honey production without treatments, you might consider managing your operation specifically toward bee production. You will have a unique product, and one that will command a premium price. Check with your state agriculture department about selling bees and queens. Some states have regulations requiring inspection, and if you ship across—or your customers drive the bees across—state lines, you definitely need to be inspected.

You can sell your bees in a number of ways, including queens, nucs, or packages.

> **Bee Smart**
>
> If you do sell bees, keep in close contact with your customers. The more successful they are, they better you look, and the more business you will have.

Education

Beekeeping education isn't usually a money maker, but it is an important way to connect meaningfully with beekeepers and the public. Schools are always looking for speakers on beekeeping and honeybee biology, but think more widely. Garden clubs, your local health food store, even blogs or podcasts are good venues.

> **Bee Aware**
>
> Many bee clubs have large study photographs that can help you illustrate a talk, available from the bee supply houses. It's much more fun to take digital photographs of your own bees and make a computer-based presentation. Many venues have their own A/V equipment, and all you need to bring is a laptop or a thumb drive.

Placing Bees

You might also find some local orchards or farms that need pollination. When it comes to agricultural pursuits, beekeeping is unique in that there is no land needed to own. Farmers, gardeners, ranchers, and individuals living in residential, urban, and rural areas all welcome bees. You have no property taxes to pay. No driveway to plow. Compared to land, bees and bee equipment are a bargain.

When you do place bees on someone else's property, you should follow a few precautions.

Keep It Clean

Before you decide to place your bees, do your homework. Even if pesticides aren't used on the land your bees are about to move to, nearby properties may not be as pristine. Your bees will fly 3 or so miles in every direction. Try to get an idea of what chemicals or pesticides may be used in the general area. It is impossible to find areas that are completely free of any toxins. However, it's good to know what you might be dealing with before you commit your bees.

Won't You Be My Neighbor?

You want to preempt any issues with the neighbors. If you are unsure of the local statutes, contact the town or city hall to make sure you are in compliance with the law regarding such issues as the number of hives allowed per acre or how close to the property line they can be, and to make them aware of you in case they do get a complaint. If your local officials know what you are up to and know you are proactive, they will help you solve problems.

Get It in Writing

When you place your hives on someone else's property, there is always the possibility there will be a problem. A neighbor might start to complain, the owner could die, an accident may happen. That's why you should always have a written agreement, no matter how casual and conversational. Whatever kind of agreement works for you and the landowner is fine. Generally, you are helping one another out. If you bring hives for temporary pollination, you should get paid for your work. A mutually beneficial permanent placement on clean, for-profit agricultural land should reasonably result in some kind of exchange of honey and produce.

Bee Smart

Most officials aren't allowed to take gifts of honey because it might be construed as a bribe, but you can always try. Neighbors will almost always accept free honey, and you should make it a common practice to sweeten your relationships with them by offering gifts of honey.

The property owner should have a copy of the agreement with your contact information on it. In addition, you should have a plan for moving the hives on short notice if necessary.

Marking Your Territory

You should mark your equipment with some sort of identifiable and permanent label. One would like to think that anyone cool enough to keep bees is cool enough not to steal your bees or equipment, but sadly that isn't so. We recommend marking your boxes and/or frames with a stencil and spray paint or some other permanent marking such as a brand. Tacking a sign or laminated note to the hive will help people find you if there is a problem, but it will not prevent theft. If the location is likely to get stumbled upon, you might also consider putting up signs warning of live, working honeybees.

Enjoying the Experience

Career tracks and financial investments that were once considered sure things are not as reliable as we once thought. But no matter how few jobs there are, honey will always have value relative to other foods. If you make less money, but spend your days in beekeeping rather than in an office, is it really less of a living? Is it really less secure?

The Beekeeping Life

Beekeeping as a lifestyle offers numerous rewards. You spend much of your time outdoors, become connected to both nature and your local agriculture, and hang out with other beekeepers. And no beer commercial has ever promised the number of girlfriends you will have in even one hive! You will feel connected with the weather, the seasons, and your entire local ecosystem in ways that few in our society have the opportunity to experience.

The Bees Themselves

In case you haven't picked up on this by now, beekeeping is an art and a science. We know no beekeeper, no matter how experienced, who isn't continually working on their technique and discovering new questions to ask. There is just something about bees—and it's more than the honey—that is endlessly fascinating. Flight patterns, behavior, stimulus/response, genetics, microbes, parasites, flowers, agriculture, woodworking, inventing, research, experiments; these are simply a few areas of observation, inquiry, and skill that we are led to by the bees. There is something about the many acting as one in simple ways that, when compounded, results in complex and productive behaviors that hold our interest and draw us in.

Sharing bees, bee stories, mistakes, and triumphs is an important part of beekeeping. Besides humans, bees have been studied more than any other creature on earth, and each beekeeper does things a little differently. There is so much to share and learn.

Bee a Mentor

Unfortunately, few readers of this book will find a mentor in their area who doesn't use treatments. However, there are beekeepers all over the world who give freely of themselves to help others succeed in keeping treatment-free bees. If you have access to a computer and are willing to spend some time online, there are several beekeeping forums where you can interact with like-minded folks. Treatment-free beekeeping conferences are springing up across the country, giving you opportunities to interact in the real world and form lasting bonds. There is also a wealth of friends and teachers, long gone now, in the authors of the great old beekeeping books from the previous two centuries. Beekeeping before the 1980s was almost universally treatment-free, and these books are fountains of inspiration and guidance.

As you develop your operation, share both your successes and your failures with others. Every beekeeper would prefer not to use treatments, and your local beekeeping community will be enhanced with the participation of members who don't use them. You, the new beekeeper, have so much to learn about bees from those with years of hands-on experience who are also eager to share. The pollination of ideas and techniques will always bear fruit.

With this, dear reader, we turn you over to the care of the bees.

May you be lured by the honey;

Embraced by the hive;

And learn from the sting.

The Least You Need to Know

- Some products require a licensed facility.
- Neighbors love honey.
- Beekeepers are as social as bees are.
- Share what you learn.

Appendix A

Glossary

abdomen The end section of a bee where the honey stomach and organs of digestion, respiration, and reproduction are located.

abscond To abandon a hive. Absconding differs from swarming in that the entire colony abandons the hive.

adhering bees Bees that are clinging to a frame when it is removed from the hive.

after swarms Swarms of bees, each including a virgin queen, that leave a parent hive in succession after the original swarm.

alarm pheromone A scent produced by honeybees that alerts other bees that there is danger.

American foulbrood (AFB) A spore-forming bacterial infection, caused by Paenibacillus larvae, which kills bee larvae.

auger hole A hole an inch to an inch and a half in diameter that can be drilled into a hive body for ventilation or entrance.

bait comb Comb from a broodnest that is moved into a new box to encourage bees to occupy the box.

bee milk Liquid food produced in the hypophrangeal and mandibular glands of young adult bees and fed to the larvae, queen, drones, and each other. Bee milk is a product of the consumption and digestion of large amounts of pollen and beebread.

bee space Space in the hive that allows bees on parallel surfaces to pass back to back without interfering with each other.

beebread Pollen fermented by the bees and used as a primary protein source for developing and young adult bees.

bottom board The bottom piece of the hive that rests between the hive stand and the bottom hive body.

break in the brood cycle Period of time when the queen ceases laying eggs. The pause in brood rearing interrupts the habitat of pests and diseases that feed and live on brood.

brood Developing bees in every stage from eggs, larvae, and pupae to fully emerged bees.

brood cycle About three weeks, the time it takes for a worker egg to mature into an adult bee.

brood food A mixture of bee milk and regurgitated crop products fed to worker and drone larvae by young adult bees.

broodnest The area in the combs where brood develops. The broodnest generally occupies the center of the hive over multiple combs.

candy A plug in the queen cage that comes in a package of bees; it must be removed by the bees before the queen can leave and join the colony.

capping A wax cover applied by the bees to encase either brood or honey.

cell One of many hexagonal tubes that make up the comb.

chalkbrood A fungal infection, caused by *Ascosphaera apis*, that kills and mummifies bee larvae.

colony A group of bees that live and work together, used interchangeably with *hive*.

comb A two-sided matrix of hexagonal cells of wax drawn by the bees.

comb guide A tool to help the bees draw comb down from a horizontal low point. A starter strip of foundation, popsicle sticks, and a wedged top bar can all function as comb guides.

complex adaptive system (CAS) A number of interrelated variables that influence one another, and respond dynamically to one another and to the environment surrounding them.

coumaphos An organophosphate that is used in some honeybee hives to control varroa mites.

critical mass The minimum number—usually six—of frames of brood a colony needs to be largely self-sufficient and to grow quickly.

crop *See honey stomach.*

crush and strain A method of obtaining honey from combs that cannot be extracted mechanically either because of comb style (top bar) or improper drawing out of the comb by the bees. Combs are crushed to expose the honey and strained to separate the combs from the honey.

cut comb Honeycomb that is cut from the frame (or top bar) and eaten without extracting.

cutout The removal and transfer of drawn comb and bees from an existing structure to a new hive. The comb is literally cut out from the structure.

dearth A period of time when forage is scarce.

deformed wing virus A common virus of the honeybee that is present in most colonies but generally doesn't display symptoms unless the hive is under stress. As the name implies, it causes the developing bees' wings to be malformed.

double screens Two screens spaced such that bees on one side can't physically touch bees on the other, preventing direct communication.

drawn comb Fully built honeycomb (as opposed to foundation).

drifting When bees enter hives other than the one they were raised in.

drone A reproductive male bee.

drone congregation area (DCA) A place where drones tend to gather (at least 30 feet in the air) and where mating generally takes place with queens from all over the area.

drone right The state of a hive being allowed to rear enough drones so that they don't feel compelled to raise them in odd places (between boxes, in honey supers).

egg First stage of bee development, lasting about three days.

embedding The process by which frame wire is adhered to wax foundation by heating the wire.

emergence When a newly mature bee leaves its cell.

EpiPen An epinephrine auto-injector used as a stop-gap response to an anaphylactic allergic reaction while medical help is sought.

European foulbrood (EFB) An infection, caused by the bacterium *mellissococcus plutonius*, that kills bee larvae.

exoskeleton The hard shell that many invertebrates (including insects) have. Literally,
a skeleton outside the body.

extracted honey Honey spun out of the honeycomb by the use of a mechanical extractor.

extractor A mechanical device that spins honey out of honeycomb by the use of centrifugal force.

feral bees Unmanaged honeybees living in the wild.

field force The bees in a colony that forage for nectar, water, propolis, and pollen.

fluvalinate Pyrethroid miticide, commonly in the form of impregnated strips, used in a treated hive for the control of mites.

food stores Honey and beebread prepared by the bees and placed in comb cells for future use.

forage Floral sources that provide pollen and nectar for the bees to collect.

foragers Bees that collect food and water.

form board A board slightly smaller than the inside dimensions of a frame used to support a sheet of foundation during the embedding process.

foundation A man-made sheet of beeswax embossed with a hexagonal comb pattern for drawing out by the bees in a frame.

foundationless Comb drawn by the bees, from the tops of frames or top bars, without the use of foundation.

frame Open rectangular housing for comb drawing (either with foundation or foundationless) that hangs parallel to other frames, bee-spaced, in a hive body or super.

fumidil A broad-spectrum antibiotic used in a treated hive for control of nosema.

grafting Moving a worker larva from a worker cell to an artificial queen cell by the beekeeper for queen-rearing purposes.

haploid clone In the case of honeybees, this is what a drone, or male bee, is. He contains half the genetic material from the queen that laid the egg, but is unfertilized.

head The top of the bee (center of sensory perception) where eyes, antennas, mouth parts, and food-producing glands are located.

hive Housing structure for the bees. Sometimes used as a synonym for *colony*.

hive body A box that holds frames for bee living and working space.

hive stand A structure that rests on the ground and supports the hive structure.

hive tool A specially designed metal device for prying hive bodies apart, removing frames, re-hammering loose nails, and whatever else may need to be done with the hive.

honey Plant nectars collected, stored, and concentrated through evaporation by the bees and used by them as their main carbohydrate source.

honey stomach The storage sac in the upper part of a bee's abdomen where nectar is carried from the flower back to the hive. Also known as the *crop*.

honey super *See super.*

Honey Super Cell (HSC) A one-piece, injection molded, polypropylene frame with fully drawn comb. The inside cell diameter is 4.9mm.

honeyflow *See nectar flow.*

hopelessly queenless A hive that has lost its queen and is lacking the resources of either eggs or young larvae to create a new queen.

house bees Bees that work inside the hive.

hypopharyngeal gland A gland, located in the head of the worker bee, which produces the substances in royal jelly and brood food.

inner cover Wood panel that fits flush with the sides of the hive and is used under a telescoping cover. Inner covers have an oval cutout in the center for hive ventilation.

large cell (LC) Worker brood cell generally larger than 4.9mm.

larva The second stage of bee development, lasting about six days.

laying queen A properly mated queen who is currently laying (or capable of laying) fertilized eggs.

laying worker A worker who, as a result of a hopelessly queenless hive, begins to lay drone eggs in a last-ditch effort to perpetuate the genetics of the hive.

locally adapted stock Bees of a genetic makeup that has been allowed to adapt to fit the local environment. Also referred to as localized stock.

mandibles Bee mouth parts used for chewing and wax formation.

mandibular glands Glands in the mouth parts of the worker bee that contribute nutrients to royal jelly and bee milk.

manipulation Anything the beekeeper does to affect the hive.

mating nuc A small nucleus colony used to mate a virgin queen and confirm that she is laying before she is moved to a different hive.

microbes Life forms including bacteria, yeasts, and molds, that are invisible to the naked eye but are visible under a microscope.

migratory cover A hive cover that fits flush with the sides of the hive.

mouse guard A barrier placed over the entrance that allows bees to pass, but is too small for a mouse.

movable comb Comb that is in frames or on top bars that can be removed individually for inspection purposes.

nectar Plant juices that are collected from inside flowers and are harvested by the bees for the production of honey.

nectar flow An abundance of nectar that allows the bees to create a honey surplus. Used interchangeably with *honeyflow*.

nosema A microsporidian infection that causes dysentery in adult bees.

nuc Short for *nucleus hive*.

nuc box A box that houses a nucleus hive.

nucleus hive A fully functioning core of a hive containing at least three to five frames of bees, brood in all stages, and food stores. A hive in miniature.

nurse bees Young adult bees, generally less than two weeks old, responsible for the care and feeding of bee larvae.

observation hive A colony that is behind glass so that the behaviors of the bees on the comb can be observed.

package Bees by the pound (generally 3–5 pounds), shipped in screened boxes with a can of syrup feed, either with an unrelated queen or without a queen.

package producer Individual beekeeper or beekeeping company who raises and prepares packages for sale. Package producers may or may not produce their own queens.

paralysis virus A virus of the honeybee that causes paralysis, quivering, and hair loss, making the infected bees look greasy.

parent hive The original hive from which a swarm issues.

parthenogenesis The process by which an unfertilized female is capable of creating fertile offspring. In bees this is known as *thelytoky*.

pheromone lure Bait from queen pheromones used to attract swarms.

pheromones Chemical signals emitted by the bees and used for communication in the hive.

pinched queen A queen that has been killed by the beekeeper by squeezing between the thumb and forefinger.

pollen Reproductive germ tissue of flowers collected and fermented by the bees as their primary protein source.

pollen flow Period of time in which there is an abundance of pollen available for the bees to build up stores.

primary swarm First swarm of the season to issue from the parent hive. The primary swarm contains about half the bees and the current queen from the parent hive.

production colony A colony used to produce honey.

propolis Plant resins collected and used by the bees in the hive for many purposes including disinfection, sealing cracks and openings, encapsulation of foreign bodies, and tuning the combs for communication.

pupa The third stage of bee development, lasting about 12 days.

pupate The transition from larva to pupa.

pyramiding up Moving brood frames from a lower box to a higher box to encourage the bees to begin working in the upper box.

queen A reproductive female bee capable of producing both drones and workers.

queen breeder An individual (or company) who produces queens primarily through grafting.

queen cage A small, screened container that holds a queen and a few attendant bees that allows other bees to feed and smell her but prevents them from killing her.

queen cell A peanut-shaped wax vessel, built by the bees, that hangs vertically on the outside of the comb for the purpose of raising a queen.

queen cup An upside-down wax cup, shaped like the cap on an acorn, that forms the beginning of a queen cell.

queen excluder A screen the dimensions of the hive body that allows workers to pass through, but is too small for the queen to pass. Used by some beekeepers to keep the queen from laying in the honey supers.

queen includer A queen excluder used for the purpose of keeping the queen in the hive as opposed to out of the honey supers.

queen right A hive possessing a queen who is capable of laying viable eggs in a healthy brood pattern.

regression Process of downsizing artificially enlarged large cell bees to naturally sized small cell bees.

reproductive split A split whose primary purpose is to increase the total number of colonies.

requeen Replacing the current queen in a hive with a new queen, either a virgin or one already mated.

resistant strains Microbial strains that have become resistant to specific treatments through selection.

reversible bottom board A bottom board that has variable depths from the bottom of the frames depending on its orientation.

rolled The injuring or killing of a queen as a result of masses of bees (including the queen) on a frame "rolling" as the frame is removed from or replaced in the hive.

royal jelly Bee milk mixed with regurgitated carbohydrates and fed to queen larvae.

sacbrood A virus-induced stress disease of the hive.

screened bottom board A bottom board consisting of a ¼" screen stretched across and attached to an open frame the dimensions of the hive body.

secondary swarm(s) A swarm(s) that issues after the primary swarm.

shotgun brood pattern A brood pattern that is scattered throughout the comb as opposed to brood that fills nearly every cell in the broodnest area.

skep Traditional straw or clay hive resembling an upside-down basket. Bees attach the combs directly to the top inner surface of the skep, requiring the removal of the combs for honey harvest and making inspection impossible.

small cell A worker brood cell, generally 4.9mm or smaller.

smoker A device with a metal burn chamber and attached bellows used for calming the bees before commencing work on the hive.

split Creating a new colony from an existing colony by moving some of the frames of bees and brood from an existing hive to an empty hive body.

starter strip A narrow strip of foundation inserted across the top length of the frame for the bees to use as a comb guide.

stimulative feeding Feeding bees to stimulate brood rearing as if a honeyflow is on.

stonebrood A fungal infection caused by *Aspergillus fumigatus* and *Aspergillus flavus*, which kills bee larvae.

super A box that holds frames for honey production by the bees, also known as a *honey super*.

supersede The act of replacing the current queen by the bees as they rear new queens.

superorganism An organism that is more than the sum of its individual components. In the case of honeybees, the organism (individual bee) is unable to function outside of the superorganism (colony).

survivor stock In the case of honeybees, bees that are able to survive and adapt without the use of treatments.

swarm A group of bees, including a laying queen and workers, that leave the original colony to establish a new colony. This represents reproduction on the colony level.

symbiote An organism that co-exists with and shares a mutually beneficial relationship with another organism.

tarsal glands Glands on the bottoms of the bees' feet that emit pheromones and sticky substances that allow the bees to climb up slippery surfaces and leave traces of their smell as they walk.

telescoping cover A hive cover that resembles a box lid. Telescoping covers hang over the edges of the hive, are a couple of inches deep, and are generally covered with aluminum sheeting.

thelytoky The process by which unfertile worker bees are able to lay fertile eggs. Bee *parthenogenesis.*

thorax The middle section of the bee (center of locomotion) from which the wings and the six legs are attached.

top bar hive A horizontal (rather than stacked) hive that uses top bars.

top bars Horizontal bars used in top bar hives from which bees draw comb down without the use of foundation or frames. Top bars butt against each other, preventing the bees from crawling up through the frames.

treatment Any substance (other than honey, pollen, or propolis) put into the hive to solve a perceived problem.

unlimited broodnest Broodnest encompassing at least three deep hive bodies (or their equivalent), in which the queen can lay without the use of a queen excluder.

varroa mite Varroa destructor, a parasitic mite of the honeybee that was introduced to the United States in the late 1980s.

worker A nonreproductive female bee.

Appendix B

Suppliers

These outlets can supply you with almost anything you need for keeping bees, often including the bees.

Shipping costs often outweigh differences in prices between suppliers. Heavy items are usually best purchased as locally as possible, or if you have a loading dock available, you can go in with other beekeepers and save significant money with truck shipments. Get a quote for the shipping if possible before ordering. This sometimes can be a sticker shock moment, as woodenware can be very expensive to ship.

Suppliers will be happy to send you a catalog if you call and request one, as they are much easier to browse than many websites. Don't be surprised (or tempted) when you see all the medicines and treatments that are sold. We only list suppliers we have done business with.

Betterbee
8 Meader Road
Greenwich, NY 12834
Phone: 1-800-632-3379
Fax: 518-692-9802
www.betterbee.com

Brushy Mountain
610 Bethany Church Road
Moravian Falls, NC 28654
Orders: 1-800-BEESWAX (233-7929)
Fax: 336-921-2681
www.brushymountainbeefarm.com

Dadant & Sons Inc.
51 South 2nd
Hamilton, Illinois 62341
Orders: 1-888-922-1293
Fax: 217-847-3660
www.dadant.com

Mann Lake Ltd.
501 1st St S
Hackensack, MN 56452-2589
Phone: 1-800-880-7694
Fax: 218-675-6156
www.mannlakeltd.com

Maxant Industries
Honey Processing Equipment
P.O. Box 454
Ayer, MA 01432
Orders: 978-772-0576
www.MaxantIndustries.com

Honey Super Cell
1102 West Andre Road
Westmorland, CA 92281
Phone: 888-343-7191
Fax: 888-396-8245
www.honeysupercell.com

Humble Abodes Inc.
636 Coopers Mills Rd
Windsor, ME 04363
Phone: 1-877-4-BEE-BOX (423-3269)
www.humbleabodesinc.com

Index

A

B

C

G

H

I

J-K

L

M

N

O

P

Q

R

S

T

U-V

W-X-Y-Z